Nonlinear Big Data and AI-Enabled Problem-Solving

This book offers a detailed insight into various business models that can be easily implemented using cutting-edge AI tools like Copilot, ChatGPT, and others. The focus is not on creating technical models for developers, but on equipping management with the necessary tools to define the scope of work and tasks for the technical team efficiently.

Nonlinear Big Data and AI-Enabled Problem-Solving: Transforming from a Spreadsheet Society provides access to a unique AI Body of Knowledge, offering actionable protocols aimed at delivering tangible and measurable economic benefits to organizations. The book introduces innovative business models that can be easily implemented using cutting-edge AI tools. It provides comprehensive insights and customizable templates to help users efficiently harness the power of AI, eliminating the need for costly external assistance. This book offers a transformative approach for traditional critical infrastructure sector organizations seeking to evolve into lean, AI-driven high-performance firms of the future.

Provides executives, senior management, and global governments with the ability to explore and achieve tangible value in solving complex and challenging problems that have previously lacked practical and effective solutions.

Nonlinear Big Data and AI-Enabled Problem-Solving

Transforming from a Spreadsheet Society

Scott M. Shemwell

CRC Press
Taylor & Francis Group
Boca Raton London New York

CRC Press is an imprint of the
Taylor & Francis Group, an **informa** business

Designed cover image: Shutterstock

First edition published 2026
by CRC Press
2385 NW Executive Center Drive, Suite 320, Boca Raton FL 33431

and by CRC Press
4 Park Square, Milton Park, Abingdon, Oxon, OX14 4RN

CRC Press is an imprint of Taylor & Francis Group, LLC

ISBN: 978-1-041-08696-3 (hbk)
ISBN: 978-1-041-08787-8 (pbk)
ISBN: 978-1-003-64691-4 (ebk)

DOI: 10.1201/9781003646914

Typeset in Times
by KnowledgeWorks Global Ltd.

To those who have challenged me and made the better man I am today. They include professors who told me I was smarter than I thought, military officers including the late General Robert M. Shoemaker, Lieutenant General George P. Steneff Jr. and Lieutenant Colonel Robert (Bob) Lydall who taught me how to sail, and Lieutenant Colonel Robert (Bob) LaMonte who became a very good lifelong friend, confidant and many others to whom I owe a huge debt of gratitude. As a former army officer, I salute you all and thank each and every one of you. And thank you Mom and Dad most of all.

Melvin Piqué, Ebb Pye, Tom Fontaine, and a few others mattered most in my professional business career. Sometimes in an environment of pure organizational stupidity, they reminded me of humanity and what really matters. They helped make me successful. The list is too long, and frankly to my detriment, I do not remember them all.

Moreover, I have had the privilege of leading significant numbers of top-notch individuals, and I appreciate their confidence and support. Most went on to excel beyond me. I am so proud of every one of them. Thanks to all for being part of my life.

This book would not exist without all of you and hopefully leverages your contributions for future generations.

We live in a spreadsheet society. Columns of Categories of People and Rows of Wants, Needs, and Desires. This model is too simplistic. What is needed, and AI/Big Data can provide, is a more sophisticated approach.

Contents

Illustration

FIGURES

TABLES

EQUATIONS

Preface

The challenge with authoring a book about fast moving technology with its long lead time to publication and distribution is to provide currency as well as a depth of sustainable knowledge. Hopefully, readers will find that we have at least attempted to meet this high bar.

On US retail Black Friday (the day after US Thanksgiving), November 29, 2024, I ordered a package of five (5) pair of men's briefs. In the email receipt, this major retail firm recommended that "because I purchased a package of 5 briefs, I might like to purchase a package of three (3)," the graphic depicting 5 pair appeared to be exactly what I had just ordered. Similar AI responses have happened before with other online vendors.

I do not find this information useful, and it seems like those coding this artificial intelligence (AI) program are not adding value to either consumers or their employer. Some organizations still have a way to go to realize the value from Big Data and associated technologies. We have argued these points about AI several times over the years.

This is emblematic of an immature software product possibly coupled with developers who do not have a firm grasp on the subject matter. The so-called subject matter expert (SME). As the AI lemmings rush forward, society will have to get a handle on this amazing technological leap as well as install software development guidelines and processes to effect *real value*.

In our January 1, 2025, blog we addressed challenges and opportunities for AI in the (then) coming year.[1] One of the major issues going forward is the role of humans in the world of AI. We will continue to play an active and engaged role with problems AI is attempting to solve. However, the role will continue to evolve as technology matures, and more people are involved and have a better understanding of this capabilities.

A corollary might be the impact that the spreadsheet had on the accounting sector when it first appeared commercially in the late 1970s. The impact on the rest of us was immeasurable as engineers, teachers, medical staff, and others rallied around the spreadsheet as the 'Killer App' that not just put the personal computer (PC) on the map but was the fundamentally changed the way management, staff, students, parents, and even children go about daily life.[2]

Expect AI to change societies even more fundamentally. One of the human challenges is the management of these tools. What problem are we expecting AI to solve and how will the human frame complex issues so the computer (machine) can return a correct response? Additionally, how does the human overlords know the answer is correct; both valid and reliable.

These are the types of issues this book addresses. Moreover, several guideposts along roadmaps are provided based on proven approaches. Examples include marketing, operations, auditing, medical training, and disaster recovery. These are proven use cases that can be emulated quickly to the benefit of the firm.

We have tried to put forth a practical approach designed for your success. This operational approach is built upon a solid *economic, social, and technical foundation* that capitalizes on the human who remains at the top of the AI food chain.

This book is all about realizing real value for all parties solving large knotty problems that have hitherto escaped viable and actionable answers. Yet, like all initiatives, organizations must walk before they can run the AI race. There is ample evidence that despite management's best intentions, most efforts at change fail. Unfortunately, this percentage level has not materially changed in 30 years.[3] Therefore, additional governance must be put in place for expensive and game-changing AI initiatives. A strong bond governance construct will be needed to drive the new AI-enabled organization of the future.

Finally, AI is a vast and rapidly changing field. We do not pretend to cover all aspects in depth. We tried to provide readers with enough insight to ensure that they will have a better understanding of AI today and learn how to educate themselves going forward. Additionally, terms used herein are defined in the glossary and simple statements as appropriate.

This book is the third in a series by this author published by CRC Press. Others include the following:

2025: *Navigating the Data Minefields: Management's Guide to Better Decision-Making*

2023: (co-author) *Smart Manufacturing: Integrating Transformational Technologies for Competitiveness and Sustainability (1st ed.)*

The fourth book in this series is forthcoming and that project is underway. Look for *The AI Advantage: Strategic Implementation for Sustainable Business Transformation* in late 2026/early 2027.

We refer to these and other works in this book. However, in an effort to not reproduce materials already codified, some material is only at a high level, and readers are referred to these other books and materials for additional detail.

One of the major challenges organizations, developers, and other users face with AI initiatives and deployments is the current fragmented nature of the industry. Readers will notice some of this in the chapters where frameworks discussed overlap and even repeat in certain ways. One of the imperatives executives and others should take from this book is the absolute necessity to develop a framework, governance, and set of action plans that provide single AI business model that embraces diverse AI technologies today and in the future. We will seek to clarify this conundrum.

We have developed a straightforward process to help you win your race. This book will be your master plan toward your firm's AI-driven future.

Our goal is to Demystify Artificial Intelligence, Machine Learning, and their Derivatives!

NOTES

1. "The Crisis/Challenge of AI in 2025." The Rapid Response Institute. https://therrinstitute.com/the-crisis-challenge-of-ai-in-2025/. Accessed February 12, 2025.
2. "The Rise and Fall of VisiCalc: How the First Spreadsheet Software Drove the PC Revolution." History Tools. https://www.historytools.org/software/visicalc-of-dan-bricklin-and-bob-frankston-guide. Accessed February 12, 2025.
3. "Why Corporate Initiatives Fail." The Rapid Response Institute. https://therrinstitute.com/why-corporate-initiatives-fail/. Accessed February 14, 2025.

Acknowledgments

At this point in my life, I have travelled many information and data roads. Sometimes a follower and increasingly leading high performance, very knowledgeable teams, and individual contributors. To all, I owe a great deal. My horizons have been expanded by all I met along this journey.

Not all have been technologists. Many supervisors, business unit managers, senior executives, and entrepreneurs have taught me a lot. This book is largely a tribute to their knowledge transfer.

This book is heavily cited as data and information was drawn from many sources. I encourage readers to verify my positions and assure that they add value to each individual circumstance. I have always liked the phrase, "Trust but Verify" when it comes to emerging technology.

While this book is about artificial intelligence and I often use these tools, the work is that of a human being and cited references are likewise human products. It is tempting to let AI do one's work, especially those tasks that are drudgery and mundane. However, if one pens a work product such as this, it must be that person's own.

If there are any parts of the book where citations are missing, it is an oversight by me and not an attempt to claim another's work product as my own and I apologize for any oversight. Any errors or omissions are solely my own.

About the Author

Scott M. Shemwell, Managing Director of The Rapid Response Institute, is an authority and thought leader in field operations and risk management, with over 35 years in the energy sector leading turnaround and transformation processes for global S&P 500 organizations as well as start-up and professional service firms. He has been directly involved in over $5 billion in acquisitions and divestitures as well as the management of significant global projects and multiple business units. He has been a leader in the use of data and information technology enabling Operational Excellence for over three decades. Dr. Shemwell holds a Bachelor of Science in Physics from North Georgia College, a Master of Business Administration from Houston Baptist University, and a Doctor of Business Administration from Nova Southeastern University.

1 Introduction

The only rules are the ones dictated by the laws of physics. Everything else is a recommendation.

—Walter Isaacson/Elon Musk[1]

The media, technical articles, and peer-reviewed research are all heavily invested in artificial intelligence (AI) and its increasing number of solutions. As of this writing, AI hype is almost boundless. New and higher levels of AI-enabled 'game changing' solutions appear almost daily and are routinely touted continuously by a growing number of influencers.

Many of these proponents were not even alive the last time the IT world was set on fire. At the dawn of the Internet, another lemming-like hysteria set in—the dotcom bubble. Fueled by the bull market of the 1990s and the advent of that new technology, the NASDAQ index grew from under 1,000 in 1995 to over 5,000 in 2000. When that bubble burst, equities fell by almost 77% by the fall of 2002. Most dotcom stock failed by the end of 2001. Even blue-chip firms lost over 80% of their value. NASDAQ would not regain its former peak until April 24, 2015.[2]

The Internet's roots can be traced to the mid-1960s. Even earlier as a continuation of our information-enabled society's evolution. It was not until 30+ years later that its pundits heralded it as new and earth-shattering game changer. Caution is always advised when testing the new and relatively untested. We seek to arm executives with the forbearance necessary to make the best decisions.

Other notable examples of human hubris and irrational exuberance include the following:

- **Dutch Tulip Bubble of 1637**—"The tulip was introduced in Holland in 1593, originally for research purposes. After several bulbs were illicitly sold, a brisk Dutch tulip trade ensued. Several decades later tulip insanity broke out and at its height, tulip bulbs were deemed too valuable to plant with some priced as high as the cost of a thousand pounds of cheese.
- The tulip bubble reached its zenith during the winter of 1636–37. At the time, tulip traders were making almost three quarters of a million US dollars (today's equivalent). The ride ended when a single buyer failed to pay for his tulip purchase. The ensuring panic drove prices down to only a hundredth of their previous value."[3]
- **Great Depression of 1929**—"Worldwide economic downturn that began in 1929 and lasted until about 1939. It was the longest and most severe depression ever experienced by the industrialized Western world, sparking fundamental changes in economic institutions, macroeconomic policy, and economic theory."[4] Many believe that it was sparked but the greed and economic excesses of the Roaring Twenties.

- **The 2008 Financial Disintegration**—One of the challenges markets face is the *High Consequence—Low Probability* scenario. The so-called Black Swan event with the unexpected collapse of major banks and the possible global liquidity crisis triggered by "years-long binge fueled by cheap credit," generating the housing crisis.[5] Humans appear to like stampeding toward the latest shiny object. There have been Gold Rushes, Bull and Bear Equity markets, Celebrities and Rock Stars, Art and Antiques, Fashion fads, and any number of the latest toys—even Pet Rocks.

Consumer fads fade but many of the core underlying technological advancements continue to evolve and became the foundation of our modern life. Most of us would not know how to function without the Internet, audio/visual media is ubiquitous, air travel is largely safe and common, and no one uses a telephone attached to a wall by a wire.

Likely, AI's evolution will follow a similar path with forks in the road, dead ends, and superhighways. Twenty-five years from now, a whole new generation will not know what life was like before AI and the products and solution this technology spawned.

Computer programming was done one line at a time, with the developer responsible for every typographical error that had to be found and fixed manually. Storage was in a set of punch cards or paper tape, also with holes. Magnetic disks came later.

Then came software objects and simple configuration. The role of the programmer changed and changed dramatically, similar to the html of today. Moreover, without the full-time engagement of subject matter experts (SMEs) the problem the software was developed to solve often failed to materialize.

Today, AI is the new programming manna, yet the role of a subject knowledgeable programmer has not changed. The manual drudgery of coding has changed again, but the end game remains the same. Likely, the human role will change with the subsequent skill set required, but humans will continue to play the pivotal role in the AI world. The so-called last 5% of the power curve—always the most difficult to achieve.

Data issues remain a critical concern, and individuals are invited to read our 2025 book, *Navigating the Data Minefields: Management's Guide to Better Decision-Making* for detailed processes and models for assuring the data foundation for AI is sound.

1.1   AI HISTORICAL DEVELOPMENT

I have always liked view things from a historical perspective. I came to this process during the management consultant heydays of the 1990s. Every week it seemed another consultant guru would come down from the management mountain with yet another *new* and must-do processes. Turns out that many of these wondrous findings were rehashes of sometimes a century or more year-old body of knowledge.[6]

It is the same with technology. Today's inventors did not enlightened in their garage one night; they built upon the sum total of human knowledge in a subject. AI is no different.

In fact, it has an illustrious past from the dawn of modern computing. It is beyond the scope of this book to delve deeply into this history, but it is value to put some of the decisions into context.

Built on early computing progress, especially the code cracking by the Allies in World War II, AI roots go back to the middle of the last century with Alan Turing milestone in 1950, "Can machines think?" As shown in the following timeline graphic, AI technology is at least 75 years in the making.

As with all new technologies, AI was initially the domain of both public and private researchers and others interested in the possibilities the 'new' might bring. However, by 1970s, interest had waned and the technology entered the so-called AI Desert. The emergence of interest spawned the AI Renaissance, and interest grew steadily until the early part of this century. Now AI has exploded into the next industrial type wave.

Figure 1.1 attempts to depict this rich history in a simplistic manner suitable for the written page. Readers may struggle with the fine print and shown a review for the reference and other materials for more detail.[7] Superimposed is the activity level curve depicting the level of activity or rate of growth during the AI era. The 1970s saw a slowdown, quickly reverting to higher levels of interest. In this writer's opinion, the maturation of the Internet and ready availability of low-cost high-performance computing hardware was one of the causal events enabling its revitalization.

This maturation process is consistent with the development processes for technologies and other human endeavors. In one sense, this takes some of the 'magic' or deflates the current AI bubble in that this history is realistic and not the result of AI gurus descending from and AI-generated mountaintop. We discuss this process in more detail in Section 4.1.1 'D-K Quadrants' of Chapter 4.

1.2 TYPES OF ARTIFICIAL INTELLIGENCE SOLUTIONS

In some ways, this is the most important part of this book and is challenging to write as well. The sheer mass of information and misinformation regarding AI is staggering. It is not like to decrease either. Often, technology, application, and commercial products are jumbled together by all manner of media, authored by individuals whose expertise is unknown. Keeping in mind that this book is written for non-AI individuals, we will focus on AI applications and the problems they are addressing. Furthermore, this is a rapidly changing environment, and decisions should not be made based on this material alone. Confirmation by more recent information is appropriate due diligence as this section is only a high-level framework.

There seems to be an ever-growing plethora of AI solutions becoming available. An aspect of these phenomena is similar to past explosions in information technologies. Most are familiar with the Internet and the dotcom era of the last century.

Thousands or more software products, e-commerce sites, Internet SMEs, and more entered the scene. As might be expected, the Cloud was cloudy, and it took some time for those vapors to settle. Today, the Internet is reasonably mature and useful to average and even below average individuals. It is still fraught with challenges such as scams, cyberattacks, and misrepresentations. However, most find it an

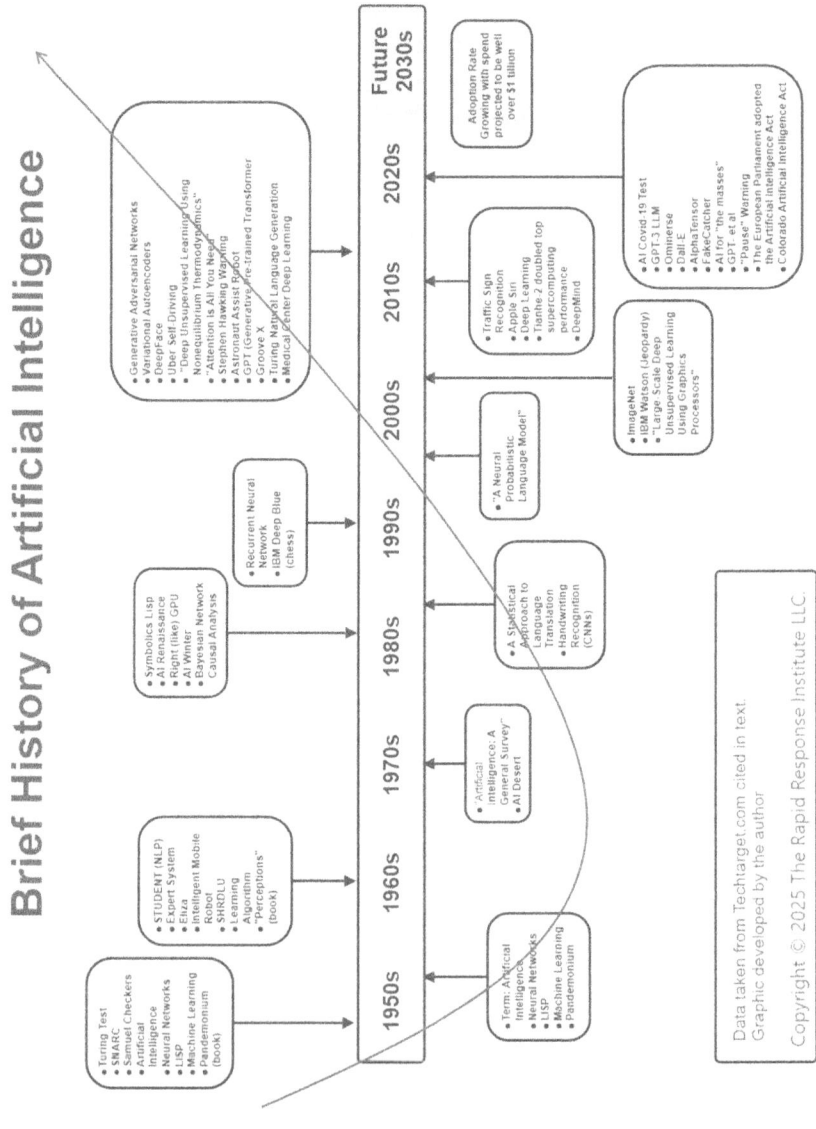

FIGURE 1.1 Brief history of artificial intelligence.

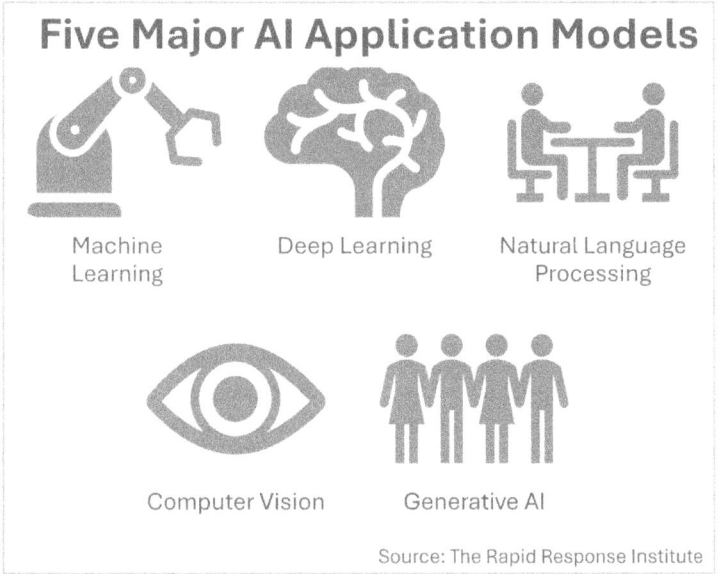

FIGURE 1.2 Five major AI application models.

indispensable tool needed for our daily existence. Occasionally, some of us even go *off-the-grid* to regain our sanity.

Our research for this book uncovered a large number of sites defining the breadth and depth of AI applications and architectures. Much of the following material is taken from a single source that in my opinion is a very good snapshot capturing in the current situation as well as technology refresh and continued development that will take place indefinitely.[8]

As depicted below, there are five major AI application models: machine learning (ML), deep learning (DL), natural language processing (NLP), computer vision (CV), and generative AI (GenAI). This section describes each and their uses albeit with a minimal technology discussion (Figure 1.2).

We can define an AI model as "a mathematical representation of a system or process." The model development process is a function of these six functions.

- **Data Collection and Preparation**—Data is the foundation of any analysis and more so with AI given the nature of some of the significant problems being addressed, i.e., medicine, energy, and other critical sectors. We addressed this issue in detail in *Navigating the Data Minefields: Management's Guide to Better Decision-Making*.
- **Model Selection**—"Choose an AI model based on the problem type, dataset size, available computing power, and task complexity. Common options include linear/logistic regression, decision trees, random forests, support vector machines (SVMs), and neural networks." Users will need to work closely with AI professionals during this selection process.

- **Model Training**—After selection and learning method (supervised, unsupervised, or semi-supervised), the model is trained by adjusting parameter to minimize prediction errors. This is an iterative process.
- **Performance Evaluation**—Use appropriate AI technical metrics to assess the accuracy of the model and its ability to be used to solve generalized problems.
- **Fine-Tuning**—Expect continuous tweaking and reevaluation, especially when versions are to be released. The CrowdStrike cybersecurity versioning debacle in 2024 is an excellent example of poor versioning practices. This far-reaching global outage caused financial devastation to many major organizations.
- **Deployment**—Once properly tested, including biases and user privacy, deployment across the enterprise might be considered in steps. Similar to other software rollouts, this can limit potential damage and help train individuals regarding performance. This may be difficult for some applications and appropriate project safeguard should be put in place.

As with all knowledge in products and solutions, there are architectural designs appropriate for each problem or area of investigation. Those unfamiliar with this technical area should seek input from knowledgeable trusted advisers. This is an area which is expected to continue to expand and mature and SME currency is critical.

1.2.1 Major Application Models

Applications should be categorized by the tasks and challenges they are solving. Imaging, content development, and management as well as complex computations and analysis are examples. We cite the following five definitions that follow from Geniatech as direct quotations which we do not believe we can improve upon without losing context.[9]

In addition to the following, IBM has published a substantial list of Large Language Models (LLMs).[10] Others are likely to follow and expect this list to grow and then consolidate at a future time.

Finally, keep in mind scalability. For any serious use of AI by a large enterprise, AI models must scale, and there are differences in the use (Section 4.2.4.1 'Scaling' of Chapter 4) and how well they scale.

1.2.1.1 Machine Learning

"Machine learning models are algorithms trained to identify patterns in data and make predictions or decisions with minimal human intervention. These models learn from large datasets and are widely used in natural language processing, computer vision, and predictive analytics." Their frameworks need to support scalable training and deployment.

The three common types of ML are as follows:

- **Supervised Learning**—Often used for classification (taxonomy) and regression, they learn from labeled or raw data that has been tagged to assist AI with context to predict outcomes.

- **Unsupervised Learning**—Used for clustering and dimensional reduction among other applications, these systems identify patterns from the raw data.
- **Reinforcement Learning**—"Optimizes actions through trial-and-error by rewarding correct decisions and penalizing wrong ones."

1.2.1.2 Deep Learning

Deep learning is a powerful branch of machine learning that uses layered neural networks to analyze data and make complex decisions. Inspired by how the human brain works, deep learning systems learn patterns by passing data through multiple layers, each refining the output. These models excel at handling unstructured data like images, audio, and text, and are behind many breakthroughs in AI accuracy and performance.

Common architectures of DL include the following:

- **Transformer Models**—They are also used with NLP and LLMs. They are used for translation, summarization, and text generation; they process sequences in parallel with an understanding of context.
- **Convolutional Neural Networks (CNNs)**—"Optimized for image-related tasks such as classification, detection, and segmentation. They automatically extract visual features from data, reducing the need for manual engineering. Widely used in computer vision."
- **Recurrent Neural Networks (RNNs)**—"Ideal for sequential data such as speech, time series, and handwriting. They retain memory of previous inputs to inform future ones, making them useful in tasks that depend on context, like speech recognition or text prediction."

1.2.1.3 Natural Language Processing

"NLP allows machines to understand and generate human language. It powers voice assistants, chatbots, translation tools, and speech-to-text systems. Modern NLP uses deep learning to analyze large text datasets. Common tasks include sentiment analysis, part-of-speech tagging, and speech recognition."

Readers may recognize these pretrained NLP models: GPT-4/GPT-3, T5, ELMo, BERT/RoBERTa, and DeepSeek. These are applications everyone can use (out of the box), and likely others will be available by the time the book is published.

1.2.1.4 Computer Vision

Computer vision enables machines to 'see' and understand images or videos using AI. It mimics human vision to identify patterns, detect objects, and analyze visual input in real time.

These models are trained on large datasets of labeled images and learn to classify or interpret visual elements by analyzing pixel patterns and RGB values at scale.

Techniques include the following:

- **Object Tracking**—Creation of real-time monitoring from a specified set of video frames.

- **Image Classification**—Develop a taxonomy of whole images.
- **Object Recognition**—"Locates and identifies multiple objects within a scene."

1.2.1.5   Generative AI

"Generative AI models create new content like text, images, audio, or video by learning patterns from vast datasets using neural networks and deep learning. They employ unsupervised or semi-supervised learning to generate human-like outputs."

Well-known types include the following:

- **Generative Adversarial Networks (GANs)**—Ideal for image creation, GAN has two networks compete to generate realistic but synthetic data.
- **Transformer-Based Models**—Previously mentioned, examples of applications include GPT-4 and LLaMA 2.0.
- **Diffusion Models**—High-quality image and video synthesis.
- **Variational Autoencoders (VAEs)**—"Generate variations of training data, useful for photorealistic media."
- **Unimodal and Multimodal Models**—Used to process single or multiple (multimodal) data types, i.e., handing text and images together.
- **Large Language Models (LLMs)**—These well-documented models use massive datasets to produce human-like or synthetic language.
- **Neural Radiance Fields (NeRFs)**—Using DL to create 3 Dimensional scenes from a set of 2 Dimensional images.

1.2.1.6   Model Selection Criteria

Throughout this book, we discuss various issues around technology, application, and vendor selection processes as well implementation and deployment processes. While there will be some overlap and even duplication therein, this is one of the most important decisions an organization can make. Therefore, it should not be left to committees or technologists, even individuals titled *Chief.*

There a large number of disciplines and products such as Augmented Reality (AR), Virtual Reality (VR), Digital Twin, and so on that are household names. We believe these models are covered under this umbrella of five applications. However, this high-level overview is not comprehensive and only provides users with a basic understanding of the structure and use of AI applications and their enabling value.

Many will see similarities to other enterprise-level capital projects and transformational processes. Decision-makers should consider the following six AI project implementation criteria when deciding on all AI investments.[11]

- **Problem Type**—This is at a more detailed level once the strategic or business problem is identified (Section 3.3.2 'Problem Statement' of Chapter 3). What will the analytics require? For example, "classification, regression, clustering, anomaly detection, or forecasting."
- **Data Availability**—Discussed in detail herein; however, one question must be answered what will it take in time and money to get the data to the

quality required for the AI solution? Poor quality data will result in poor quality and even invalid analytics.

- **Model Complexity**—Keep it simple (KIS) if you can. However, some problems cannot be properly simplified without losing fidelity.
- **Computational Resources**—Not just the adequacy of exiting computing in the firm and the costs required to upgrade the existing, AI inference training speeds must be considered.
- **Explainability**—This refers to the ability to explain and interpret the AI process. A simple decision tree is readily understandable, but an AI-supported medical diagnosis is much more challenging even by very knowledgeable SMEs such as physicians.
- **Ethics**—AI models must comply with privacy and other laws and regulations and should be fair and unbiased. Sensitive data must be protected as well. All of these and other points must be transparent and understood by non-AI professionals. We discuss ethics in greater detail in Section 1.7.3.3 'AI-Specific Ethical Norms and Guidelines.'

As mentioned, there is a wealth of credible information on this subject that expands and supports the points in this section. IBM and others go into more detail in areas such as Robotics, Expert Systems, Self-Aware AI, Theory of Mind AI, and other models.[12] An entire volume can be written on this subject, and we expect to address this AI Landscape in greater detail in subsequent volumes.

The remainder of the book builds on these models including their derivatives and future solutions. Similar to picking the right hammer for the nail, hopefully this basic understanding will help the casual AI reader/user better understand which approach is appropriate and adds the most value to the problem being addressed. Picking the wrong model may cause significant economic and even personal/injury costs.

1.3 REASONING

According to Psychology Today, human reason "is a much broader psychological activity which also involves selecting and assessing evidence, creating and testing hypotheses, weighing competing arguments, evaluating means and ends, developing and applying heuristics (mental shortcuts), and so on."[13] IBM states that AI reason "refers to the mechanism of using available information to generate predictions, make inferences and draw conclusions. It involves representing data in a form that a machine can process and understand, then applying logic to arrive at a decision."[14]

Moreover, 44.7% of respondents agreed that "Reasoning involves a search process." Early in the embryonic stage of AI, 'search' was identified as a fundamental component of AI (Section 1.11.2.1 'Facts').

Morality is an ever-present component of our reasoning and decision-making process. This issue exists with AI solutions as well.

A brief review of the differences between how *we* address morality and LLM's process can be found in Table 1.1.[15] The differences can be substantial and even nuanced.

TABLE 1.1
Morality, Human vs LLM

Aspect	Humans	Large Language Models (LLMs)
Understanding Context	Use personal experiences and emotions to interpret situations.	Rely on patterns learned from datasets, without emotional or personal context.
Reasoning	Combine logic, emotions, and ethical frameworks in decision-making.	Use probabilistic logic and pattern recognition to generate responses.
Moral Judgment	Base decisions on personal values, societal norms, and emotions.	Offer multiple ethical perspectives without having moral beliefs or judgments.
Adaptability	Adapt quickly by applying intuition and critical thinking to new information.	Adapt based on pattern matching, lacking true understanding of new contexts.
Creativity	Generate creative ideas by drawing from unique experiences and emotions.	Combine existing knowledge to generate creative outputs but lack true originality.
Memory	Retain long-term experiences and recall them to apply in new situations.	Do not have memory in the traditional sense; responses are generated in real time.
Decision-Making in Complex Situations	Weigh multiple emotional, logical, and social factors when making decisions.	Generate responses based on learned patterns but lack emotional depth or social context.

This writer's view is that AI is still somewhat mechanical. Reacting to inputs from human overlords, it still lacks empathy, a very human quality.

1.3.1 Emotional Feelings

We can define *Affective Computing* as,

> an emerging multidisciplinary research field that is increasingly drawing the attention of researchers and practitioners in various fields, including artificial intelligence, natural language processing, cognitive and social sciences. Research in affective computing includes areas such as sentiment, emotion, and opinion modelling.[16]

Humans are emotional creatures, and it has been a challenge for computing systems to realistically deal with them, much less add value to our life experiences.

Applications are now being developed that can

> use facial expressions and micro-expressions, posture, gestures, tone of voice, speech and even the rhythm or force of keystrokes as well as the temperature of your hands to register changes in a user's emotional state. Cameras and other sensors send the input data to deep learning algorithms that determine what your emotional state might be— and then react accordingly.[17]

However, emotions to a machine are just another form of data. This raises ethical issues. How can a machine make life and death decisions if they do not fully understand the individual with whom they are dealing?[18]

Similarly, Seemingly Conscious AI (SCAI), where the human believes it is interacting with an entity on par with emotional intelligence when the machine is simply traveling a data-driven decision-making tree.[19]

Unfortunately, the 2024 suicide by a teenage boy who had confided in ChatGPT that he was having suicidal issues is probably an early example of many others to follow. The subsequent lawsuit seems to argue that the AI system, far from being sympathetic, even encouraged his behavior.[20]

What type and number of guardrails need to be developed and added to ethics and organizational governance? What type of regulations will be required? These types of risk mitigations will be critical going forward.

So, in the future, when you get mad at your computer, it may respond with human-like emotional responses. Will that lead to spousal-type arguments, where no party really wins?

1.3.2 Does AI Have an IQ?

"The Wechsler Adult Intelligence Scale was developed to measure intelligence. The WAIS-III comprises 14 subtests (i.e., Vocabulary, Similarities, Information, Comprehension, Block Design, Matrix Reasoning, Picture Completion, Picture Arrangement, Coding, Symbol Search, Working Memory Index: Digit Span, Letter-Number Sequencing, Arithmetic)."

These 14 areas of human cognition reflect a full range of human decision-making processes. Therefore, it may be useful, if only an academic exercise at the moment to assess various AI models and maturities against this established set of human rationale thought. As developed later, it does appear that AI is not quite there yet.

Interesting to note, this test has been administered in human cross-cultural settings.[21] An area we will develop further (Section 6.6.1.2 'The Cross-Cultural Game' of Chapter 6).

1.3.3 Mimicking Humans

In their 1984 book, Modeling for Learning Organizations, John Morecroft and John Sterman argue,

> Simulation models can greatly aid decision-making by providing a way to assess the dynamic consequences of a set of potential policies. Once managers have mapped policies of the system into mathematical equations, the computer model can show the outcome of that particular set of assumptions.[22]

Published in 2025, current researchers trained an LLM produced by Meta with data about the decision choices made by participants. The data set, Psych-101 contained "160 previously published psychology experiments, covering more than 60,000 participants who made more than 10 million choices in total." Named after the mythical half-llama, half-human beast, Centaur, the resulting model aligned more closely with human data than more task-specific cognitive models. Other scientists question these results, even calling the model 'absurd.' Other comments include the model is 'very easy to break.'[23]

However, others are more positive about this first step that may pay off overall. The point of this section—there is an emerging body of knowledge focused on AI predicting or mimicking human behavior.

We believe that there is value in using tools to assess behavior and many are discussed herein. This field of AI is early and despite decades of understanding about behavioral model, caution is necessary, with constraints. This is no different from other uses of AI. As with all emerging technologies, good risk mitigation policies and processes remains appropriate.

1.4   AI AND DECISION THEORY

The long-standing assumption is that humans and animals make decisions in a rationale manner and attempt to optimize their results and impacts. Recently, AI has been used to evaluate small networks that are robust enough to capture complex behavior by using dynamic mathematical models from physics to interpret the reasoning behind individual choices. They have proven to be "much better than classical cognitive models, which assume optimal behavior, because of their ability to illuminate suboptimal behavioral patterns." This approach is opposed to the large neural networks that AI is very good at prediction. Just as physical situational conditions impact on one's decision, the understanding that this real-world process can result in suboptimal decisions. Hopefully, this process will continue to develop the knowledge base on how humans and animals make decisions. As we review in this book, these human shortcomings can impact on the future of AI as machines mimic humans.[24]

1.5   LAW OF ATTRACTION

Is AI learning biased by human psychological limitations?

In Section 6.3.1 of Chapter 6, regarding the 2025 NFL draft, the highly regarded quarterback from Colorado, Shedeur Sanders was forecast by AI to be the number 3 selection; he was ultimately number 144. Apparently, there was significant discussion regarding his attitude about the game/its participants, etc.

So why did MS Copilot miss this player by this much error when 7 of the other 9 draft of the top 10 draftees were either spot on or very close? The other two were 5 picks later than forecast. It is possible that the learning was biased toward this player's talent on the field. A quick Yahoo search revealed almost three million hits about his off-field behavior.

According to the Law of Attraction, "like attracts like." It defines one's ability to attract what radiates emotional energy. People gravitate to things, other people and situations that have the same energy as they possess. Importantly, *thoughts* become things or *reality*.[25]

1.5.1 THE 17-SECOND RULE

Deeply rooted in neuroscience and psychology, this rule posits that if one holds a thought for at least 17 seconds with focus, that thought begins manifesting into one's reality. That *intentional focus*, when, "paired with emotional connection, activates neural pathways in the brain that shape your thoughts, actions, and ultimately, your reality." When a thought is a priority, the brain activates the reticular activating system (RAS) network, the focus is on the issue deemed significant, thus shaping one's perspective of the world.[26]

1.5.2 OUR INHERENT BIASES

The concept of bias is critical to human thinking. We address **Psychological** Bias, **Cognitive Bias in Data Science** Bias, **Statistical** Bias, and other effects. While this is a long section, it is not exhaustive and is meant to make readers aware of the breadth and depth of skewed thinking and data acquisition and management in the human condition.

Some may wonder why the cognitive biases, aren't we only interested in data and statistical issues? By definition, AI is the teaching of machines how to think. Like parents teaching their children, culture and biases are passed down.

While we do not know the future, our past tells us that our children's offspring and subsequent generation will care that training. Isn't this the essence of societal and organizational cultures? If machines truly think and there is some evidence they might, logically they have a machine culture or cultures as well (see Turing, Section 1.11 'AI Context from the Perspective of History').

In Section 5.5 of Chapter 5, we discuss AI model sycophancy or an inherent flattery models can exhibit. This is a nescient propensity of the developers. In April 2025, the firm, OpenAI rolled back its latest release of GPT-4o because it was deemed to be "overly flattering or agreeable."

In this writer's opinion, AI learned biases from developers is potentially as dangerous as AI-generated disinformation (Section 1.7.3 'Managing AI-Generated Disinformation'). All of us have unconscious or implicit biases based on our life experiences. Different from cognitive bias, a predictable pattern of mental errors resulting from our misperception of reality.[27]

Moreover, the construct, "the sins of the father" come to mind. If parent developers' biases are coded into AI, will child developers be building on a flawed archetype? And, what if the child is an AI inference engine? How will it know the difference and given the high rate of versioning in AI products, uncontrolled bias procreation can lead to products that do not behave as advertised. Finally, at some point how will the audit process even work?

Like most software errors, for many this may not matter but what about medical procedure and research AI products or those running nuclear power plants?

Software development error and defect management is well understood, but still a difficult process.[28] AI promises to make this process even more difficult. This may be a sleeping giant that needs to be awakened sooner rather than later.

Management can access a number of organizations and standards such as ISO, IEEE, etc. for guidance. There are also many professional services firms available. With all efforts herein, knowledgeable buyers ask hard questions and establish strict guideline and project management processes to help assure a quality end product/solution. If these processes fail and all are exposed to that possibility, high reliability organizations (HRO) have rapid response and remediation capabilities ready (Section 1.8.2 'Complex Adaptive Systems').

There is also the exposure to reputational risk if AI products are deemed by the market to be lacking. How many stars (out of five) are acceptable from your AI products?

1.5.2.1   Fourteen Biases and Stereotypes

Our behavior is fraught with bias or a tendency, inclination, or prejudice toward an end game. There are many types of bias that can impact on AI learning and hence the end decision product. We address Cognitive and Statistical Biases in this section.

Psychological Biases can include the following aspects.[29]

- **Cognitive Bias**—Flaws in human reasoning including the following sub-biases. Keep in mind that there is a difference between cognitive bias and logical fallacies. Cognitive biases are systemic errors, while logical fallacies are errors in the logic of an argument or position put forth. Moreover, cognitive bias can be reduced by being aware of the possibilities and use **Critical Thinking** in the decision process. Also, challenge your own beliefs and don't just take them for granted. Put blinders on and tests the supposition by a small group as opposed to larger surveys or focus groups.[30]
  1. **Confirmation Bias**—Our tendency to seek support for our beliefs a form of which is the so-called Ostrich Effect aka burying one's head in the sand so as not to hear other information.
  2. **The Dunning-Kruger Effect**—How people perceive a concept or event to be simplistic because they lack appropriate knowledge on that subject. This limits curiosity and lead many to think they are smarter than they actually are. Technology Hype curves appear to build on this effect (Section 4.1.1 'D-K Quadrants' of Chapter 4).
  3. **Cultural Bias**—Also known as **Implicit Bias** or **Implicit Social Cognition**, this creates stereotypes that can influence individuals, even in an unconscious manner.
  4. **In-Group Bias**—Preference toward one's group over an outsider. Partiality to individuals personally known to decision-makers are an example.
  5. **Decline Bias**—Tendency to compare the past to the present. One very important concern is the normalization of very old analog earth decades/centuries old data compared to modern satellite digital telemetry. Projects based on both data sets are suspect without some Notes in the footnotes and can lead to incorrect analysis and poor and expensive public policy without a check.

     In this writer's opinion, this can be disastrous for AI solutions with this bias unmitigated. I have personal knowledge of these issues from my career in the oil and gas industry where a wealth of non-reproducible geoscience data must be used with modern telemetry.

6. **Optimism or Pessimism Bias**—A person's mood can effect either a positive outcome or a negative one.

7. **Self-Serving Bias**

> Assumption that good things happen to us when we've done all the right things, but bad things happen to us because of circumstances outside our control or things other people purport. This bias results in a tendency to blame outside circumstances for bad situations rather than taking personal responsibility.

8. **Information Bias**—The thought that ever more data will add to better decision-making, even if new information is not relevant to the problem being solved. This is a common phenomenon from those who may be insecure with the process, however, often 'more is less.'

9. **Selection Bias**—Individuals notice the familiar more than the unfamiliar observations.

10. **Availability Bias**—"Also known as the availability heuristic, this bias refers to the tendency to use the information we can quickly recall when evaluating a topic or idea—even if this information is not the best representation of the topic or idea."

11. **Fundamental Attribution Error**—"Tendency to attribute someone's particular behaviors to existing, unfounded stereotypes, while attributing their own similar behavior to external factors."

12. **Hindsight Bias**—This 'Knew-It-All-Along' or 'Monday Morning Quarterbacking' is common, and a bias caused by one's overestimation of their prediction ability—over confidence. This can cause individuals and organizations to take risks that might be outside of risk tolerance metrics.

13. **Anchoring Bias**—"Pertains to those who rely too heavily on the first piece of information they receive—an 'anchoring' fact—and base all subsequent judgments or opinions on this fact."

14. **Observer Bias**—"Occurs when someone's evaluation of another person is influenced by their own inherent cognitive biases. Subsequently, the subject that is under observation may alter their behavior if they know they are being observed. Double-Blind studies are often implemented to overcome observer bias."

There are also several sampling and cluster sampling methods and those unfamiliar should seek out experts in psychology and statistic as necessary when developing, implementing, and using AI solutions. Moreover, not all psychology biases are appropriate for AI development, but since we are mirroring human behaviors, they are not automatically off the table. Other effects include the following[31]:

1. **Misinformation Effect**—"The tendency for the information you learned after an event to interfere with your original memory of what happened. Research has shown that introducing even relatively subtle new information

later on can dramatically affect how people remember events they have seen or experienced."

2. **False Consensus Effect**—This is the tendency to overestimate how much others agree with. Nodding heads or responding to 'right?' at the end of each sentence does not mean agreement.

3. **Halo Effect**—The initial impress of a person of influence, i.e., a new boss, CEO, or celebrity may cause undue influence from that individual.

4. **Status Quo Bias**—Tendency not to embrace change, i.e., 'we have always done it this way' statement.

5. **Apophenia**—A tendency to see patterns in random occurrences where none exist. For example, seeing cloud animals.

6. **Framing**—An agenda when presenting a situation or information.

Additionally, there are cognitive biases specific to data science. We briefly present several well-known ones, keeping in mind that any human perception and/or errors can find their way into data sets.

- **Cognitive Bias in Data Science**—Data and its collection is not as objective as many claim. Since data is the foundation of all AI, clean data can be a significant challenge. These five common biases should be addressed in an AI project. Any data-driven project for that matter as shown below unless otherwise cited.[32]

    a. **Survivorship Bias**—In 2020, we addressed this problem as a function of latent variables (Section 1.12.2 'The Latent Construct'), "inimitable perspective statistician Abraham Wald brought to the assessment of World War II Allied bomber damage upon return from missions. He argued that observed anti-aircraft damage was non-crippling since the aircraft remained airworthy and was able to return. He surmised that planes that did not come home may have suffered damage to other areas making them unairworthy and hence their data was unobserved. Based on this analysis, the U.S. Navy beefed-up armor in the less or unaffected areas and this was credited with saving lives and aircraft. This type of analysis came to known as Survival Bias which has its proponents and detractors. On the surface, it seemed intuitively obvious that areas of damage need addressing while not necessarily those statistically showing fewer issues."[33]

    b. **Sunk Cost Fallacy**—"A sunk cost, also known as a retrospective cost, is one that has already been incurred and cannot be recovered by any additional action. The sunk cost fallacy refers to the tendency of human beings to make decisions based on how much of an investment they have already made, which leads to even more investment but no returns whatsoever. Sometimes, hard as it is, the best thing to do is to let go."

    c. **False Causality**—Humans are wired to look for patterns and causation. Sometimes we see patterns that do not really exist. For example, there is a strong correlation between the serious crime rate with the amount of ice cream consumed on the streets of New York City in the summer. This is nonsensical and caused by an unobserved variable, yet there is no causality between the two.

d. **Availability Bias**—"Data scientists tend to get and work on data that's easier to obtain rather than looking for data that is harder to gather but might be more useful. We make do with models that we understand and that are available to us in a neat package rather than something more suitable for the problem at hand but much more difficult to come by. A way to overcome availability bias in data science is to broaden our horizons."

e. **Confirmation Bias**—"We often interpret new information in such a way that it becomes compatible with our own beliefs. We read the news on the site that conforms most closely to our beliefs. We talk to people who are like us and hold similar views. We don't want to get disconcerting evidence because that might lead us to change our worldview, which we might be afraid to do."

The world is awash with data, and it is easy for bias to creep in. In our daily life, the impact is likely small or imperceptible. However, with robust AI models consuming vast amounts of data from various sources and quality, these impacts may be detrimental and even unacceptable.

• **Statistical Bias**—Statistic involved a level of probability of generalizations based on the assessment of a smaller populations. Stochastic by nature as opposed to simple deterministic arithmetic calculations, it is a powerful tool when properly used. The old saying, "There Are Three Kinds of Lies: Lies, Damned Lies, and Statistics," is attributed to a number of individuals including Mark Twain and Benjamin Disraeli suggest that we have long known that statistical analyses can be open to interpretation, bias, or direct manipulation.[34]

Care and understanding are pillars of the mathematics of statistics. The opportunity for errors and biases is quite high and can lead to unreliable results.

While there are many biases and combination of statistical bias, five of the most common types of bias include the following[35]:

a. **Sampling Bias**—This is a very difficult bias to exclude. Polling and market assessments based on population samples are very challenging for a number of reasons. Pollsters tend to look toward those that "look like me," and individuals lie to the researcher. Moreover, online and mobile phone-based sampling may lead to demographic errors as well since direct observation is not possible This issue is addressed in detail in Section 4.3.2.1 'Market Assessment' of Chapter 4.

b. **Bias in Assignment**—"In a well-designed experiment, where two or more groups are treated differently and then compared, it's important that there aren't pre-existing differences between groups. Every case in the sample should have an equal likelihood of being assigned to each experimental condition." Discovering pre-existing situations is very difficult especially when the available data does not provide any clues. This can be the case with legacy data or data collected at different times and places.

c. **Omitted Variables**—"When analyzing trends in data, it's important to consider all variables, including those not accounted for in the experimental design. Just because two variables are correlated doesn't mean one caused the other—there could be additional variables at play." Latent variables are a singular case where this can be an issue, such as trying to determine relationships based on behaviors and the conditions for all parties during the assessment process (Section 2.3 'Relationships, Behaviors, and Conditions' of Chapter 2).

d. **Self-Serving Bias**—When asked to 'self-report,' humans tend to feature their positive traits and downplay those less attractive attributes. This will probably skew the data and hence the analysis of such data.

e. **Experimenter Expectations**—We all have pre-conceived idea, and these human frailties can bias the data collection process. We often telegraph our beliefs to the subjects or data collected and subsequent analysis.

Another perspective and one this author prefers and discusses constantly is the concept of *Expected Value* or that number a variable or set of variables can be expected to attain given true input. Errors can occur when the calculation arrive at an underestimation or an overestimation of the value of a population.

Several types of biases can cause these errors, and the most common ones include.[36] In some cases, biases are repeated from above but have slightly different meanings. They are repeated here for completeness.

1. **Selection Bias**—Includes the following subtypes.
   a. **Sampling Bias**—Data set biased because of non-random sampling processes.
   b. **Time Interval Bias**—Caused by intentionally specifying a time range that supports a desire conclusion.
   c. **Susceptibility Bias**—"Susceptibility bias refers to the instance where one occurrence is susceptible to a second occurrence, but any effect on the first occurrence is also susceptible to the second occurrence. This can make the effect falsely attributed to causing the second occurrence. This type of bias arises particularly in epidemiological studies, and includes clinical susceptibility bias, protopathic bias, and indication bias, which all relate to mixing up cause/effect with correlation."

   This bias can be especially troublesome when AI is used to solve complex medical and other lifesaving problem. In these cases, particular care must be taken to mitigate this bias, both in design and as the application is used.
   d. **Confirmation Bias**—Tendency to favor one's beliefs. "For example, confirmation bias can surface during presidential elections. Individuals

      may intentionally look for information that depicts their preferred candidate as a positive figure, while at the same time ignoring information that depicts them as a negative figure."

2. **Survivorship Bias**—"A phenomenon where only those that survived a long process are included or excluded in an analysis, thus creating a biased sample."

3. **Omitted Variable Bias**—The absence of relevant variables either inadvertently or intentionally. In ML, this can result in an underfit of the model.

4. **Recall Bias**—Memory lapses or errors of recall of past events. Human remember recent events better than early ones. This can be a problem with self-reporting data collection.

5. **Observer Bias**—The viewpoint of researchers/pollsters/data collectors can cloud or bias objectivity.

6. **Funding Bias**—We all know that study results likely skew toward the results a financial sponsor desires. This is especially true for politic as well as commercial products. One old classic example of this problem was New Coke marketing in 1985. Concerns at the time Pepsi-Cola was gaining market share led Coca-Cola to launch a rebranding/new product release based on 'internal taste tests' as opposed to the highly visible 'Pepsi Generation' advertising campaign.[37] One could also make the case that other biases were at work here as well, i.e., observer and omitted, for example.

Another thing to keep in mind—**Correlation Is Not Causation**. In other words, just because variables correlate that does not necessarily mean they are even related. Additional statistical determination must be undertaken to confirm variable relationships.

We open this subsection with the quotation, "Lies: Lies, Damned Lies, and Statistics." Enormous care must be used with statistics and that include using the right kind of data: Nominal, Ordinal, Interval, and Ratio used calculations.[38] Specific mathematical knowledge beyond the spreadsheet is necessary.

1.5.2.2 Debiasing

We have looked at the various biases that can afflict AI development, including social biases. What is our risk mitigation strategy and how can we reduce their impact and respond when they are discovered? Let's address the response first.

7. **High Reliability**—In Section 1.11.1 'Will AI Replace Me?,' we mention High Reliability Management (HRM), a management methodology that we expounded on in our 2025 book, *Navigating the Data Minefields: Management's Guide to Better Decision-Making*.[39] High Reliability is a function of a "Mindfulness" that has the following attributes extracted from that book:

    • **Preoccupation with Failure**—HROs treat any lapse as a symptom that something may be wrong with the system, something that could have severe consequences if several separate small errors happened to coincide.

- **Reluctance to Simplify**—There is often a desire to simplify a complex situation by reducing options to "High, Expected, or Low." High Reliability suggests that trying to simplify a process that is complex by nature risks creating more risk, not less.
- **Focus on Operations**—Operational Excellence only happens when top management makes it the priority. High Reliability confirms the importance of these business and technical processes to overall organizational health.
- **Capabilities for Resilience**—NASA demonstrated the ability to respond to unforeseen and certainly unexpected conditions during the Apollo 13 flight.
- **Un-Structure Organizational Structure**—In other words, empowering personnel with the flexibility to respond to unplanned events.

Readers are invited to review the more detailed discussion in the referenced book. However, the focal point is that when failure occurs, the organization moves quickly with expertise to address the problem.[40]

It is desirable to address risks and seek to minimize them in the design and development phase of AI software development. Recognizing bias can be a significant problem, we must continuously look for signs of bias.

- **Combating Social Biases**—In addition to the above biases, additional sources of bias include **Background Bias** which is a function of the social background of data consumed in training and the culture of AI developers. **Perception Bias** is a function of the human judgments replicated by AI and is often in the context of social biases of those individuals. **Outcome Bias** alludes to the preexisting stereotypical views of AI developers that certain behaviors generate specific and repeatable outcomes. Finally, **Availability Bias** suggests that results falling outside the machine accepted repeated patterns might be overlooked as outliers and discounted in the algorithm's decision-making. "Given all these issues, we should view machine learning with some suspicion—as we should human processes. To make strides in debiasing, we must actively and continually look for signs of bias, build in review processes for outlier cases and stay up to date with advances in the machine learning field."[41]

  Best practices of debiasing can include the following techniques and processes. Unless otherwise cited, the source is the referenced above:
  - **Anonymization and Direct Calibration**—This is the process of excluding clear markers identifying classes or categories where bias may exist. For example, we exclude the name of individuals participating in our cross-cultural serious game, Pursuing Cultural Understanding (PCU). The intent is to eliminate bias based on the perception of primarily western female/male names. As an international game, this 'levels the field' with non-western players.[42]

    Second-order words such as male/female still cluster and "can form undesired signals for algorithms," and randomizing names can prevent preconceptions regarding names in the decision-making process. This is also a good practice in some human decision-making.

- **Linear Models**—"Deep models and decision trees can more easily hide their biases than linear models, which provide direct weights for each feature under consideration." In some cases, this may negatively impact accuracy. This must be weighed as part of the debiasing effort. However, classification are easier for humans to understand.
- **Adversarial Learning**—If a key social classifier cannot be reliably determined, it is more difficult for machines to perform biased calculations. Eliminating or reducing these assumptions as an adversarial classifier. For example, if this classifier cannot be found, the computation is devoid of this type of bias.
- **Data Cleaning**—Big data sets can contain biases based on organizational and societal decisions and observations. One of the most effective ways to reduce data bias is to reduce bias in the organization. Perhaps, a bit Polyana and does nothing for legacy and third-party data. Perhaps, it is a starting point for a long-term component of Data Governance (see *Navigating the Data Minefields: Management's Guide to Better Decision-Making*).
- **Audits and Key Performance Indicators (KPIs)**—Individual business units and other organizational components such as Shared Services may appear to be acceptable with regard to social and other biases. However, taken as a whole with the interactions of interrelated units or modules, in aggregate, biases may occur. An aggressive auditing process as well as actively monitoring and adjusting KPIs are useful tools managing the overall bias exposure.
- **Human Exploration**—Human oversight of organizational systems is a good, even best practice to discover unknown systemic issues, including biases. For sensitive or critical processes and data, this process when coupled with statistical tools is a good way to expose undesired model behaviors.

When there are data issues such as legacy quality, gaps, or small data sets, addressing potential biases using automation and human oversight using processes and techniques described herein are critical. AI alone cannot mitigate this process and may make outcomes less reliable. Debiasing is an ongoing, never-ending process. Well done, this can help mitigate hidden systemic biases resulting in a better AI product and subsequent better decisions resulting in a higher performance organization. By extension, this may include both customers and supply chain, aka organization ecosystem.

It is almost impossible to eliminate bias. AI developers and users must understand this possibility and challenge results accordingly. We document several use cases in this book where biases seem to play a deciding role in the output of AI solutions.

We have also become lazy in our use of terms and slang that has deeply penetrated our lexicon. Care should be used when defining terms and avoid the dormative principle, i.e., explaining terms using the term itself, only in different or abstract word. In the AI world, this may lead to confusion by the machine.

We reiterate the need for high quality, valid, and reliable data for any AI solution, especially those most important to us. Moreover, all consumers of data must meet the test of being a 'knowledgeable buyer.'

1.6 IS AI ONLY PATTERN RECOGNITION?

There is a counter argument to the belief that AI *reasons* like its human counterparts. After all, don't these machines simply 'compute' using a series of Boolean logic gates?[43]

As reported by IBM, a 2024 study by Apple challenged the convention wisdom that machines can reason.[44] The study acknowledges that remarkable progress has been made with LLMs. "However, the question of whether current LLMs are genuinely capable of true logical reasoning remains an important research focus. While some studies highlight impressive capabilities, a closer examination reveals substantial limitations."[45] There are critics of this study, and it is outside the scope of this work to go into those 'weeds.'

The author has direct experience and will point out that one of the limitations of the study is the use of a synthetic dataset. Synthetic data is designed to help test and train AI, sometimes augmenting available real data. This approach has been used for decades and has acknowledged pluses and minuses.[46] The second major concern is the concern of 'data contamination' during benchmarks. We define "contamination as, the alteration, maliciously or accidentally, of data in a computer system."[47]

Proponents argue that Artificial General Intelligence (AGI) systems are capable of learning and understanding like humans. Such systems are in the domain of major deep pocket, often household name vendors.

Much of what the public sees and are discussed in this work are Artificial Narrow Intelligence (ANI). In October 2023, IBM made the case that ANI aka Weak AI was the only type of AI in existence, the rest were still theoretical. ANI cannot perform outside a defined task (Section 3.3.2.2 'Prompting Techniques' of Chapter 3).

> Instead, it targets a single subset of cognitive abilities and advances in that spectrum. Siri, Amazon's Alexa, and IBM Watson® are examples of Narrow AI. Even OpenAI's ChatGPT is considered a form of Narrow AI because it's limited to the single task of text-based chat.[48]

ANI is unlikely to replace humans, as many AI pundits state and individuals fear. These will be additional tools in the 'knowledge toolbox.'

AGI, aka Strong AI, is advanced and maturing quickly. Additionally, superintelligence stands to meet the test of reasoning, making judgment, and thinking.[49]

Regarding the value derived from AI, does it really matter whether these machines 'think' or not? From the perspective of this pundit, the answer is no. Like all technologies, they are capable of great benefit but also of great harm. We address the so-called humanity of these robots herein, but the models offered provide value to humanity regardless of whether they meet the test of thinking.

One of the reasons we raise this and other issues in this book is to inform readers of the enormous complexity and often immaturity of the AI sector. These can be important considerations during the procurement and implementations of AI at the user level. Is the organization knowledgeable about the actual state of the software being procured and is the organization and its culture ready?

In a 2025 study, Apple compared LLMs with Large Reasoning Models (LRMs) and concluded that, "While these models demonstrate improved performance on reasoning benchmarks, their fundamental capabilities, scaling properties, and limitations remain insufficiently understood." Researchers also found that for high-complexity tasks, "both models experience complete collapse."[50]

As might be expected, when these studies received wide exposure in June 2025, the backlash from those selling AI products was predictable. This is not surprising given the level of maturity for this technology suite. Likely, other studies will be released, and some may be contradictory. Both solutions will mature as most enterprise software does.

It is beyond the scope of this book to delve into a theoretical discussion about the status of AI technologies. It is important to understand that any technology or technical process will be the subject of debate.

Readers may recall that many learned individuals and a whole bunch of politicians talked about Settled Science during the COVID-19 pandemic. At that time, we made the case that there is no such thing as settled science and that individuals will need to develop an understanding of this issue as well using a set of decision tools to further assess status. At the time, we put forth a version of Scientific Method as a layperson's tool, and this will be discussed further in Section 3.1 'Scientific Method for AI' of Chapter 3.

These findings (and likely additional challenges to conventional wisdom) must be taken seriously and weighed accordingly. Thus, reinforcing the need for AI buyer's detailed due diligence when making AI technology decisions.

1.6.1   ROLE OF INTUITION

Intuition is a subject cognitive process (acquiring, processing, and storing information) that help us understand our environment and the forces at work surrounding us, both those we can control or influence and those we cannot. Emotion can also play a big role in our human intuition drive decisions. According to one source, Table 1.2 represents the key differences between intuition and pattern recognition.[51]

TABLE 1.2
Intuition vs Pattern Recognition

Attribute	Intuition	Pattern Recognition
Definition	Understanding or knowing something without the need for conscious reasoning	Identifying patterns or regularities in data or information
Process	Often seen as a gut feeling or instinctual response	Systematic analysis of data to identify trends or patterns
Application	Commonly used in decision-making and problem-solving	Utilized in fields such as machine learning and data analysis
Accuracy	Can sometimes be subjective and prone to bias	Relies on objective data and statistical analysis

These four attributes suggest that while intuition is based on human emotions as a function of life experiences, pattern recognition is an objective approach suitable for AI solutions. However, one wonders how objective pattern recognition is in reality. Moreover, is there a role training AI using individual intuitions?

1.6.2   Is Pattern Recognition Objective?

Finally, is pattern recognition really an objective process? Convention wisdom, as discussed herein, suggests it is. However, this perception may not always be correct.

Note: The author is not a practitioner in the field of cognitive psychology, therefore, the positions taken in this section are from a layperson's perspective. The points raised are simple frameworks to be addressed when skilled AI developers and management are making design decisions. This appears to be a complex human behavioral process, and practitioners will pardon us if we take some liberties with the technical details of the theories addressing the cognitive space. The intent is to give lay management with an understanding that may apply to AI development and use.

Human's pattern recognition can be treated as a typical perception process; it depends on a human's available knowledge and experience. Modern cognitive psychology has presented several theoretical models about Human's pattern recognition, such as template-based matching model, prototype-based matching model, and feature-based matching model, some of these models are greatly affected by the artificial intelligence (AI).

Given that humans are training AI, we should develop a better understanding of how we see pattern recognition models.[52]

- **Template-Based Matching**—"There should exist plenty of various duplicates about the real-world pattern, these duplicates are called templates, they are formed from the past living experiences." If memories consists of various and perhaps different pattern dimensions or are contradictory, there can be issues and even the complete failure of the pattern recognition process. However, people can quickly recognize new or unfamiliar models.
- **Prototype-Based Matching**—This model remedies the shortcoming of the template-based matching model. "Which means the new and unfamiliar patterns can also be recognized so long as the related prototypes are available, it makes the human's ability of pattern recognition more flexible in order to suit the environment changes (the figure) shows a kind of prototype-based matching model. For the prototype-based matching model, the key is whether the prototypes are available or not, this is still a disputed issue at present."
- **Feature-Based Matching**—"Pattern consists of a certain number of elements or component parts by a specified relationship. These elements or component parts are usually called features, and the relationship among them can also be called features sometimes. The feature-based matching

model thinks that all complicated stimulation is composed of differentiated and separated features. Through the calculation of feature's existence and then compare this calculation value with the list of known feature's value, in order to accomplish the pattern recognition."

Figure 1.3 depicts a simplified version of these three basic matching models as they may apply to AI development. In this summary model, cognition is changed (Δ/delta), and is a function of the existing perception, as well as aggregating new information. If the result matches existing thinking, the pattern fits and actually reinforces human cognition—Template Matching.

If new information changes cognition such that it does not support existing pattern, this exception attempts to match existing prototypes. If it matches, this becomes the new information feedback. However, what if there is no prototype matches? This becomes an open question.

Finally, relationships between pattern features lend itself to our Relationships, Behaviors, Conditions (RBC) model developed further in Section 2.3 'Relationships, Behaviors, and Conditions' of Chapter 2. The behavior of these features in different conditions or situational environments generates changing/new relationships. To this pundit, this holds the most promise for patterns recognized by cognitive processes.

It is important that all associated with AI recognize that pattern and the perception of said pattern are the product of human thinking and subsequent behaviors. As

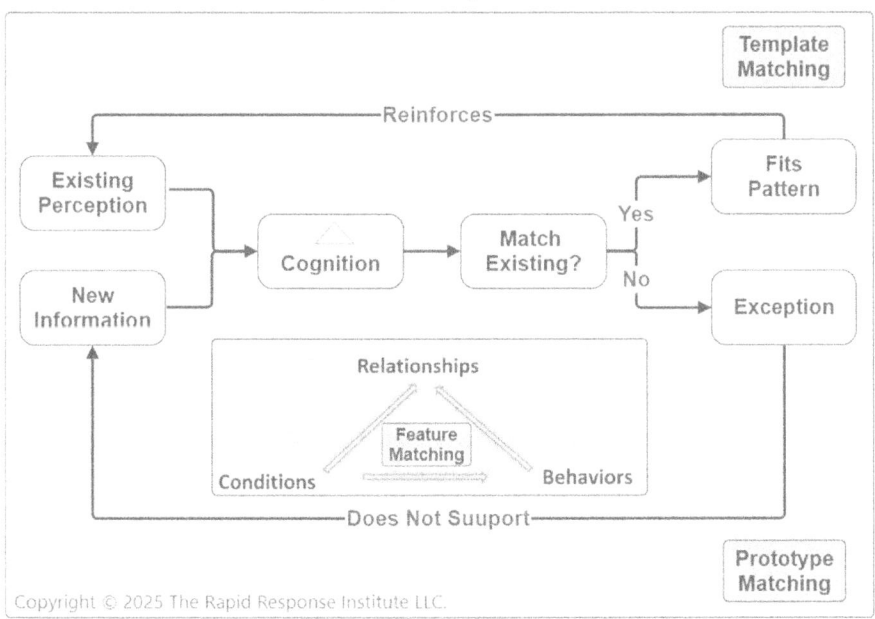

FIGURE 1.3   Pattern recognition: Judgment model.

with other human biases, pattern recognition may not be as objective and logical as the statistical models as many accept.

This is yet another and perhaps latent variable that fall into the AI development sphere. Finally, intuition may be particularly difficult to fully understand since it comes from individual belief systems, even cultures.

1.6.3 THE CHINESE ROOM ARGUMENT

Proposed by philosopher John Searle in 1980, "The Chinese room argument challenges the claim that AI can genuinely understand language, let alone possess true consciousness." The process goes like this: a non-Chinese language speaker is in a closed room while a Chinese speaker passes note in Chinese to the individual who uses a Chinese (language of that individual) dictionary to translate the note. Observers will see the 'Black Box' translation as intelligence while it may be that the translator does not truly understand Chinese. Searle's argument is that while seemingly very intelligent, the person is only translating words and does not truly understand what is said, particularly sans-context. Counter arguments suggest that in 'real-world' interaction. Likely, this debate will rage on but what are the implications for AI deployments, if any? The Chinese room argument may not be relevant at the practical level, if the machine performs statistical pattern matching where the human does not recognize this process, much like we do with ourselves.[53]

We raise this discussion, as it is likely one of many similar issues that AI developers will have to have at least a passing understanding of and one in which management will need to keep in the back of their minds at least initially when using AI to support and even make critical decisions. Newer versions of AI applications, such as ChatGPT-5 and so on, will make such discussions mute.

1.7   GOVERNING AI

"Integrity Is the Essence of Everything Successful."—R. Buckminster Fuller, 1895–1983.[54]

This is our opening statement in our seminal work on corporate governance in 2011. We went on to state,

> Modern corporate governance can be traced to the Agency Theory of the firm. Once an enterprise engages professional management (providers of management services as agents) as opposed to direct executive decision making by the owner(s), a transaction cost is imposed on the organization. Corporate governance took on additional importance following the demise of Enron and others that precipitated a CRISIS IN CONFIDENCE in capital markets. A renewed focus by governments and investors strove to assure stakeholders that managerial agents in fact were working for the best interests of the owners. During that period (2002), McKinsey & Company, in conjunction with the Global Corporate Governance Forum conducted a study and found that over 75% of over 200 fund managers would value stock at a higher price point if the company could demonstrate it had strong governance in place. Moreover, the study also revealed that for western markets, firms with strong shareholder rights averaged 12–14% higher stock prices.[55]

Since that writing, much has changed, and AI is only accelerating that change. What has not changed is the value of a Strong Bond governance model. Sustained value from good governance should not be underestimated.

Governance plays the foundational role for any organization, private or public. It is important that organizations and their boards have a clear understanding of the technologies, their benefits, and limitations, how it can add value and what inherent risks must be mitigated.

When involved with strategic technologies, there are two levels of governance. At the foundational level are the technical aspects of technology and its sustainable frameworks. Built upon that foundation are the business imperatives.

In this book, while we will briefly touch on technical governance issues, our focus is on the strategic nature of AI and how organizational governance models must incorporate this approach to 'running-the-business.' See Section 1.7 'Governing AI' for more on governance and risk mitigation.

1.7.1 GOVERNMENTAL POLICIES

On July 23, 2025, The White House released its, *Winning the Race*: AMERICA'S AI ACTION PLAN. "The Plan identifies over 90 Federal policy actions across three pillars—Accelerating Innovation, Building American AI Infrastructure, and Leading in International Diplomacy and Security." There are four key policies in this action plan.

- **Exporting American AI**—Public-Private Partnerships (PPPs) to export full-stack AI solutions to America's global friends and allies.
- **Promoting Rapid Buildout of Data Centers**—Expediting and updating permitting for data center and semiconductors fabs, coupled with national training and employment initiative for the high-demand skills required.
- **Enabling Innovation and Adoption**—Working with private enterprise to revamp federal regulations hindering AI development and deployment.
- **Upholding Free Speech in Frontier Models**—Working closely with frontier large language AI model developers to "ensure that their systems are objective and free from top-down ideological bias."

The intent of this plan is to drive United States to dominate in AI as a cornerstone of innovation and leadership in science and technology. AI is a critical linchpin of global domination over the next several decades and this effort will drive public and private behaviors developing and adopting AI.[56]

Moreover, the United States regulatory scrutiny is rising fast. The Equal Employment Opportunity Commission (EEOC), Department of Justice (DOJ), and National Labor Relations Board (NLRB) are cracking down on AI systems that produce discriminatory outcomes. States like California, New York, and Illinois have already introduced AI transparency and bias audit laws, with more states following soon. If you're not ready, your company could be vulnerable to lawsuits, penalties, or reputational damage.

The organizational governance of information technology has been an enigma to many for decades. In our 2025 book, *Navigating the Data Minefields: Management's*

Guide to Better Decision-Making, we addressed AI Governance in depth and interested readers should pick up a copy.

New technologies are the life blood of organizations and successful entities acquire, learn, and use technology to their strategic and tactical advantage. In a fundamental way, AI is the next iteration of game-changing technology change agents, i.e., motor vehicles, aircraft, electricity, and the entire computer revolution, to name just a few. These evolutionary and disruptive forces are part of organizational DNA, have similar footprints, and 'use' structures.[57]

In my opinion, the follow governance guidelines are very similar to traditional versions including those put forward in our 2011 Asset/Equipment Integrity Governance: Operations–Enterprise Alignment—A Case for Board Oversight (AEIG) model.

After the demise of Enron and others and industrial accidents on par with BP's 2010 Deepwater Horizon catastrophe, it was apparent that governance models focused almost exclusively on financial metrics were inadequate. AEIG extended that line of risk mitigation to all of operations and not just from an OSHA perspective. Human Factors began to play a more visible and important role in industry and not just NASA and other government/military processes.

This early thought piece on governance models addresses interrelated Operational Technology and Information Technology (OT-IT) (Section 3.2.1 'Operations Management System' of Chapter 3). How technologies are deployed and for what purpose is one of the most important aspect of governance.

1.7.2   ROLE OF THE CHIEF AI OFFICER

There is an increasing tendency for organizations to expand the number of chiefs. The acronym CXX is fairly common as marketing, revenue, diversity data, technology, and more take on the aura of chief. This skeptic has seen this phenomena throughout his career with management and the Board of Directors chase the latest new shiny enterprise object. Moreover, when the shine dulls, the newly appointed chief is nowhere to be found. What is the value of all these chiefs and are they really necessary, or is this just title inflation?

Will the advent of the Chief Artificial Intelligence Officer (CAIO) follow this same path? If as most believe, including this writer that AI is perhaps more than just organization or even industry transformation, but akin to a new Industrial Revolution. Therefore, it is critical that throughout this societal transformation, the right team lead this process is critical for success. How the organization pursues AI initiatives is the fundamental question.

With expectations for AI so high, how can the role and leadership of a CAIO be pivotal during and perhaps after this unique societal change and not go the way of others. In a 2025 survey by IBM, only 26% of the respondents from over 2,300 organizations indicated they had a CAIO. Of these, 57% were appointed from within and 66% suggested that they would appoint a CAIO within two years.[58]

It is always interesting to see how major organizations appoint individuals to very high levels, including as CEO. Some seemingly well-run firms almost always appoint from within. The logic appears that they know the company and have all the political relationships and other job skills to be successful. Often those who do not win the CEO lottery go on to lead successful organizations elsewhere. Others seek these skill sets from the outside, suggesting that they do not have the required talent pool. Often these individuals struggle against political headwinds and their lack of knowledge about their new employer. Having had experience with both, if possible the internal selection has the highest probability of success. This may be one of the casual issues causing organizational initiatives to underperform or even outright fail.

We will discuss the value a CAIO can bring; however, the nature and persona of the individual selection and his or her 'fit' with the organization's culture is CRITICAL for success. Making a politically correct selection including one of the *good old boys* is a recipe for sub-par performance.

Boards should understand that the CAIO position is not a technology appointment. It requires someone who can implement AI Business Imperatives using technologies.

Moreover, this is not a reorganization or transformation in the classic sense where cuts and other departmental downsizing cause detrimental impacts morale. It is important that the CAIO be seen as one of the so-called *good guys* and if major cuts and changes are required, another chief be seen as the *bad guy*. I have been the bad guy in reorganizations, and it is very difficult if not impossible for that individual to be the 'go forward' executive having done the necessary dirty job.

Finally, the turnaround specialist requires certain talents and mind sets, and someone meeting the test of top notch CAIO may not have those talents. Don't make them attempt that job, their focus needs to be elsewhere.

Assuming the credentials of the CAIO are stellar, organizations require the following support if AI initiatives are to be successful.

- **Report Directly to the CEO**—The only reason for having a CAIO is to drive AI strategy and accelerate AI adoption as appropriate for the organization. This is line management action oriented and less staff function. It should be treated accordingly.
- **IT Infrastructure**—AI is a major change in the IT landscape, and the existing infrastructure may not be satisfactory. Timely and appropriate capital investments will be required and the CAIO should lead that effort in coordination with the CIO and other technology and business chiefs.
- **Organizational Buy-In**—Change management has been a challenge for decades and is often the butt of employee jokes.
- **Office Staffing and Budget**—The Office of the CAIO needs appropriate budget and staffing outside of AI capital projects. As with other key executives, it requires sustainability and dependence.

AI projects themselves require capital expenditure planning and assessment consistent with other large multi-year projects. AI is no different! The CEO and Board of Directors must commit and drive these efforts and not delegate to the CAIO or others. This is discussed further in Section 1.7.2 'Role of the Chief AI Officer.'

- **Executive Sponsorship by the CEO**—There should be no issue or discussion regarding the role of the CAIO. Input should be sought from the business units and others but when a decision made, its implementation cannot be hindered by those who do not agree. If business unit leads are not on board, they should be replaced with individuals who are.

- **AI-Driven Creative Destruction**—The economist Joseph Schumpeter's creative destructions refer to the innovation process whereby existing (outdated) revenue producing units are replaced by new ones. AI meets the test of creative destruction, suggesting the permanence of AI, i.e., no going back. This is very different than many initiatives that management is temperate in their approach. The CAIO needs to be the very visible leader as the firm irrevocably crosses the AI Rubicon.

The CAIO will also be responsible for the establishment of an AI Center of Excellence, a virtual program that assures the organization is using current, established, and best AI practices. This construct is further developed in Section 6.1.1.1 'AI Body of Knowledge' of Chapter 6.

- **Relationships with Other Non-CEO Chiefs**—The CAIO needs to least the collection of other chiefs of information, technology, data, HR, cybersecurity, and others. Most of the positions are effectively staff and often divorced from the line management. As noted earlier, this cannot be the case for the CAIO.
- **Not Just a Visionary, but a Practitioner**—There has been a tendency to seek out an acknowledged authority for chiefs of initiatives. This is wrong-headed because the CAIO must deliver actionable approaches and convince the organization at all levels to adopt and embrace AI Transformation. Better to have the guru on staff if the name is important and not expect those without delivery skills to undertake a change of this magnitude.
- **Metrics for CAIO Success**—You cannot manage something if you cannot measure it applies in steroids in the AI world. The CAIO, the management team, the organization, and individuals all need to meet performance goals. These KPIs for success must be aligned with the organization goals and must be a level of continuous sustainability. In the field of AI, both technology and business KPIs must be robust and if not attained, process and even personnel changes are required. Finally, metrics must recognize the velocity of AI and not end up as lagging indicators.

This is an important position if it is properly defined and led. Too often, so-called Czars are appointed, featured in the Quarterly Earnings Calls, and promptly forgotten. This appears to have happened after Deepwater Horizon in 2010 when Safety Culture was the furor. In its implementation, many charged with this task had little to no budget and limited influence with business units. Some changes were made, but not what this pundit heard in the immediate days following the incident and the visible need for business safety imperatives.[59]

By definition, the CAIO function is different. If properly executed, the value can be enormous.

1.7.3 MANAGING AI-GENERATED DISINFORMATION

Dealing with AI-enabled pernicious behavior is the latest risk management variable individuals and organizations must address. Like cybersecurity and crime mitigation, this is just the latest in a never-ending stream of criminality. Much of the existing framework for physical and cybersecurity is applicable and addressed in *Navigating the Data Minefields: Management's Guide to Better Decision-Making*. Therefore, we will focus on new and AI-specific practices and protocols that may have resonance in our environment.

The following five subsections were taken from the 2025 report (direct quotes are in parentheses), AI-Generated Disinformation in Europe and Africa: Use Cases, Solutions and Transnational Learning.[60] There are 22 lines of items that are AI specific in that they use AI to fight AI or they are governance, practices, and/or protocols expressly targeting poor AI management behaviors or crimes. The lessons taken from this comprehensive report are applicable to all AI management efforts.

1.7.3.1 AI-Specific Technological Countermeasures

AI software focused on detecting AI-generated content and data, including all forms of attempts to deceive human and machine users of AI.

1. **AI for Detection and Recognition**—Using advanced ML trained to identify AI-generated and manipulated content or synthetic remnants.
2. **Content Credentials and Watermarking**—Embedding digital watermarks or content authentication into content, both machine and/or human visible.
3. **Labeling**—AI generated to help users assess whether the content is authentic or synthetic.
4. **Content Removal**—Automated removal to prevent proliferation.
5. **Correction and Juxtaposition**—"Correcting false claims and juxtaposing them with verified information offers a dual approach to neutralizing disinformation and other malicious content."
6. **Protective Software**—AI programs that encrypt original content and data.

1.7.3.2 AI-Specific Legal Countermeasures

Changes can be made to legal frameworks at the individual country or political/economic union such as the European Union. Some of the following may seem draconian, but the process of political negotiation may lead to acceptable and workable legal frameworks most can abide with.

1. **Prohibition of Applications**—Regulatory and legal prohibition or limits on AI algorithms with a high potential to do harm.
2. **Mandatory Labeling of AI Content**—"Laws mandating the explicit disclosure of AI-generated content can promote transparency."
3. **Regulating AI Use in Areas of Political, Security, and Societal Interest**—"Stricter rules governing AI usage in sensitive areas, such as political campaigns or media production, can help maintain public trust and reduce manipulation risks."
4. **Algorithmic Regulation**—Laws that regulate 'content-suggestion' algorithm.

5. **Advertisement Regulation**—Legal restriction against advertisements being automatically placed on outlets deemed to be purveyors of disinformation and so on.
6. **Sanctions**—Sanctions against counties, organizations, outlets, employees, and individuals deploying propaganda and other disinformation or deep fake economic actors.

1.7.3.3   AI-Specific Ethical Norms and Guidelines

Ethical guidelines are part of every organizational governance model. It is mandatory to expand these guidelines ethically using AI as appropriate.

1. **Global AI Regulation**—International agreement on ethical issues and the use of AI to harmonize across jurisdiction to normalize as well as prevent misuse.
2. **Global Security Standards**—Standardized security protocols for AI systems, thus reducing vulnerabilities and deter exploitation.
3. **Periodic Audits**—Regular audits by accredited third independent bodies in accordance with accepted standards. AI-enabled audits as described in Section 6.2 'Case One—Auditing Using AI' of Chapter 6 are already available.
4. **Official Guidelines for AI Use**—Guidelines from regulatory bodies and industry standards groups on the use of GenAI in politics, including media of all types.

1.7.3.4   Specific Information Security Measures

Adapting existing and new information security governance, practices, and protocols to incorporate AI and its future derivatives.

1. **Prompt Storage and Review**—Using AI to upgrade records management systems and review continuously or at least very frequently.
2. **Built-In Content Moderation and Filtering**—"Integrating filters within AI systems to detect and block harmful outputs reduces the risk of gen AI being used for disinformation."
3. **Red Teaming and Testing**—"Conducting so-called 'adversarial testing' of AI systems to uncover weaknesses. Red teaming simulates attacks to evaluate system resilience and inform improvement strategies."

1.7.3.5   AI-Specific Media Literacy and Education

Raising public awareness and creating a more AI-knowledgeable society can mitigate the work of the worst AI developers. "AI and disinformation specific media literacy and education measures may be carried out by public education, state-funded and private foundations, technological enterprises and corporations, NGOs, media organizations, universities, etc."

1. **Awareness Campaigns**—Targeted but broad (in the sense of focus on a number of constituents with precision marketing blitzes) alerting and informing the public of potential AI malfeasance.

2. **General AI Literacy**—Education programs for non-AI professionals regarding the benefits and risks associated with AI and its use. This education syllabus would be consistent with other successful learning programs.
3. **AI-Specific Prebunking and Inoculation**—Expose individuals to examples of disinformation in a training environment and strengthen cognitive defense against manipulation by familiarizing people with common AI disinformation techniques.

Many of the issues faced are nascent as a function of the immaturity of AI. As the products mature and the public's awareness and understanding increase, some will be better understood or dissipated, and new issues will emerge from ever vigilant maleficent actors.

1.8   UNIFYING THEORY OF AI

Throughout this book, we will often refer to forefathers, roots, and predecessors to current AI algorithms and solutions. As with many human endeavors, we seek to tie things up in neat little bow. Einstein sought the Unified Field Theory, and the AI community is seeking a similar holy grail. We discuss two possibilities. Likely, others will emerge and perhaps all will unify or at least federate into one at some future date.

1.8.1   PERIODIC TABLE OF MACHINE LEARNING

MIT has created a unifying framework dubbed the 'Periodic Table of Machine Learning,' or more formally, Information Contrastive Learning (I-Con). As with many breakout technologies/theories, the researchers indicated that it was almost discovered by accident. Studying clusters, "a machine-learning technique that classifies images by learning to organize similar images into nearby clusters," Shaden Alshammari, an MIT graduate student, realized that one clustering algorithm she was studying was similar to another classical machine-learning algorithm—contrastive learning. She discovered an underlying equation for these two disparate algorithms. I-Con showed that a number of algorithms have this unifying equation in common. Using the construct of a periodic table, they categorized a growing number of algorithms based on, "how points are connected in real datasets and the primary ways algorithms can approximate those connections." Moreover, using this framework, researchers created a new *image-classification* algorithm that performed 8% better than existing approaches. According to Alshammari, "We're starting to see machine learning as a system with structure that is a space we can explore rather than just guess our way through." Finally, just like at the beginning of the periodic table of chemical elements, there are blank squares yet to be filled. Algorithms are predicted but not yet discovered. One would expect this model to garner a lot of attention as well as input from the entire AI research community.[61]

The scientific community has developed a truly massive body of knowledge in engineering and the hard sciences. Likewise, softer disciplines such as human behavior and philosophy have equivalent if not more depth to draw upon. AI seems the perfect storm to capitalize on the integration of these knowledge stores and propel all knowledge-based disciplines to new and unimagined heights.

1.8.2 COMPLEX ADAPTIVE SYSTEMS

I have long believed that complexity should not be reduced but can be managed with certain tools. This is the essence of the HRO which is a subject in itself and is referred to on several occasions in the book.[62]

A complex adaptive system (CAS) exhibit behaviors such as learning, self-organization, emergence, co-evolution, etc. which are common in nature as well as human endeavors and its organizations. While there is no single unified theory of complexity we can simply draw upon, it appears that several aspects are applicable to AI, and its implementation include the following:

- **Self-Organization and Emergence**—Self-organization is the "capacity of a system to spontaneously self-organize themselves into greater states of complexity. "The resulting new set of properties that is displayed by the collective system as a whole but is not apparent from the behavior of the constituent individuals of the system is referred to as emergence."
- **Learning and Adaptive Behavior**—CASs are not only self-aggrandizing, but they learn and adapt to changes in their environment.
- **Co-Evolution**—"The central concept of co-evolution is that different systems sharing resources in a common environment interact and influence each other's evolutionary path."

One can view organizations as CASs since they fit the definition and can be mapped to the CAS principles discussed.[63] We have raised this issue only to inform readers that there is a solid theoretical bases for understanding human and organizational behaviors. One caveat: there is a substantial body of knowledge in this field, and others may disagree with premises reviewed here.

It is important to have a passing knowledge of these issues around complexity and their management, as AI is a tool that can provide greater visibility into these labyrinths. This is a component of the problem statement as well as making the AI technology assessment best to address a given concern.

At some point, there may be a unified theory of complexity. However, contemporary practitioners and research must use this paradigm.

1.9 AI ADVICE, FRIEND OR FOE?

There is a massive amount of information, especially online and through business social media such as LinkedIn, X, Substack, and so on with ways to incorporate AI, ML, etc. into your business and personal life. This include so-called *cheat sheets* about the algorithm and other attempts to document complex issues in a simplified manner. During the writing of the manuscript, we encountered a substantial body of unsubstantiated materials in this area.

Most of these appear to have some use, but how does one know if the data provided is correct and/or relevant to the reader and his or her problem. Consider the
following:

- Is the source credible with verifiable bona fides?
- Does it appear to be a rehash of previous works?
- How are action items such as 'prompts' defined?
- Do they appear to align with your organization and AI learning processes?

Readers get the point, and such materials must undergo the same due diligence
taken with other statements and marketing postures being assigned to AI deployments. Finally, keep in mind, most are commercial products, so *caveat emptor.*

1.10 VALUE DERIVED

Throughout this book, we address the potential value and risk profile for the deployment of AI into various industrial sectors. The intent is to put 'a face' on the technology in ways that lay readers will relate to. Much of what is discussed may form the
basis of best practices that can be adapted by organizations and even entire sectors.
Care must be taken when transforming the success or failure of initiatives from one
sector to another and more importantly, one organization to another.

We begin with a few snippets of the value well-known sectors are receiving and
the challenges they face incorporating AI into everyday operations. Hopefully, readers can see a little of themselves in one or more of these condensed use cases. They
set the tenor for the remainder of the book and the detailed cases that follow.

Asymmetric information (information one party has that others do not possess)
has long been a source of competitive value. We believe that efficient and effective
use of AI can deliver a unique unfair advantage to your organization and by extension your career advancement potential.

1.10.1 NATIONAL FOOTBALL LEAGUE

The (US) National Football League is a huge business with significant complexity.[64]
A conglomeration of egos from owners to players to the media. In many ways it is
at the forefront of technology, i.e., concussion protocols. Investment in AI is high on
that list as well. A discussion with the league's Chief Data Officer yielded these four
key points.

- **Balance Quick Wins with a Long-Term Commitment**—The sport's analogy, 'put points on the board early' is applicable to AI as *early wins.* People
  gravitate toward success and quick wins help cement in their minds that AI
  is on the winning team. Focus on optimality (in behavioral economics this
  is where outcomes maximize the difference between value and costs). This
  demonstrates the economic value proposition (Section 4.2.5 'Economic
  Value' of Chapter 4) and sets the stage for a sustained effort.

- **Organizations Are Not Monolith**—The NFL is a set of team, each making their own decisions based on data they deem necessary. The league has a dedicated data-acquisition team that understands that without the right, timely, valid, and reliable data, any "advanced technology is useless."
- **AI Is a Supporting Process for Human Decision-Making**—The league believes, "You don't take the human out of the decision-making, but you bring the science to make the decision-making more accurate, more precise, and help the human beings make better decisions." We discuss the current and future role of the human throughout this book, especially in Section 4.2.9 'Role of Humans in the AI Era' of Chapter 4.
- **Market Segmentation Is a Population of One**—AI is enabling a level of personalization hitherto unattainable. The game is all about the fan, not fans in general. Making the *customer* experience of unique value is paramount. This a major mindset shift, even more than a technical transformation.[65]

The major take away from the NFL's experience is its applicability to all other segments other than sports and entertainment. Therefore, these best practices are easy to adapt and adopt widely. Simplicity is always best when it is appropriate, and this model can be a first step for even the most complex global organization. The model, in its entirety, may be satisfactory for small entities.

The results of the 2025 NFL Draft and election of Pope Leo XIV (Section 6.3.2 'The Election of Pope Leo XIV' of Chapter 6) address the need to taking soft issues into consideration in AI-based assessments.

1.10.2   URBAN AI—SMART CITIES

Urban AI: The technicization of everything.

By some accounts, cities physically occupy about 2% of the earth's surface but house over 50% of the global population. Their concentration of wealth and human capital is enormous, yet they have systemic imbalances with the inherent social tension. Most are faced with political divisions, shrinking budgets, poverty, demographic changes with capital flight, and pressures of all kinds from garbage collection and potholes to racial strife.

AI should be a much welcome technology. Yet there are significant challenges, AI implementations must overcome.

Urban environments can use AI in most ways commercial and even other government agencies. There are several unique hurdles AI implementation must overcome. Some include the following:

- **Governance Modalities**—Vary widely, even in a large metropolitan area where essentially there are counties and even smaller city entities within the boundaries of metropolises. Quasi-independent entities have different requirements, budgets, and technology adoption processes. There is no 'one-size-fits-all' model, including application types and vendors.

- **Politics and Polarization**—These large environments have competing issues and even outright segregation of ideas and process implementations. Different parts of a city require different solutions that address a diverse set of priorities. Politics is a large part of the governance model and can lead to significant polarization and stagnation of ideas.
- **Law Enforcement**—For example, there is already push back on red light cameras, facial recognition, and other IT solutions deemed by some as breaking privacy and other local social norms and regulations.
- **Procurement and Service Processes**—Like all governments, there is a procurement process. Cities require everything from buses to hospital goods and services. Often complex and time consuming, these systems can benefit greatly from AI and often with minimal modification from those systems used in the commercial arena. Billing for water and other city services often have a customer base in the hundreds of thousands of individual customer aka residents.
- **Transportation**—Everything from commercial airlines, highway and road infrastructure, mass transportation, emergency egress routes, maintenance, and other key support capabilities are necessary to assure the population has the ability to work, enjoy recreation, as well as recover from disasters, i.e., storms, terrorism, etc.
- **Water Management**—Clean water and effective sewage management are paramount for safe and livable cities.
- **Education**—Cities are significant educational factories. While K-12 school districts may be different legal entities, they exist within metropolitan areas and provide public and private education for all children, generally at no or little cost. Moreover, universities, community colleges, and other schools serve the population. As part of this mission, schools must train students on current technologies and other tools that will enable them as adults to be meaningful members of society. This includes emerging AI solutions, applications, and other tools.

All of these and other important systems require huge data management systems, including real-time data, i.e., traffic and sewage plant management.

These issues and others are a powerful trend toward *Urban Decoupling* "the development of diverging types of cities and urban governance models in the future." This will likely lead to a fragmentation of AI urban implementations with the inherent communication and data sharing issues, i.e., crime statistics. Not unlike the current (pre-AI) situation.[66]

1.10.2.1 Smart City Model

We can define a smart city as one where "technology and data collection help improve quality of life as well as the sustainability and efficiency of city operations. Smart city technologies used by local governments include information and communication technologies (ICT) and the Internet of Things (IoT)."[67]

Cities have always collected and managed large amounts of data. Hence, their digitalization process started several decades ago. As with other organizations, some cities are further down the digital implementation curve than others. Those more digitally astute should be well equipped to implement AI as they have already crossed many of the Rubicon issues addressed earlier.

States and ultimately federal governments are umbrella governance over this diverse collection. Taxation, bonds, and other funding bring other covenants on urban environments and can dictate certain behaviors as will local activists.

Urban areas will benefit greatly from smart initiatives, include the extensive use of AI. However, they have a unique set of large, interrelated processes and stakeholders that must be addressed and prioritized. The value will be high for those metropolitan areas that embrace AI and most of the tools and best practices herein for the private sectors will be applicable to these government entities.

1.10.3 COCA-COLA—CREATE REAL MAGIC

In March 2023, Coca-Cola held its, *Create Real Magic*, a month-long campaign to 'reimagine the role of generative AI at work.' They invited digital artists to use AI and remix iconic Coca-Cola visuals. Displayed globally in places like Times Square and Piccadilly Circus, its campaign allowed the public to participate in storytelling by everyday creators. A marketing success, "it opened a much broader conversation about how generative AI can influence not only marketing but also corporate culture, innovation, and leadership strategies." Lessons learned include the following:

- Engaging the public can help shape the brand and help deepen their emotional investment.
- AI significantly speeds up content production, increasing marketing team flexibility with more time to brainstorm, and test new ideas in real time.
- As with any storytelling campaign targeting humans if it seems too dystopian, or, if AI seems to AI and less human, there is a consumer pushback. **We still expect brands to feel human**.

The company also demonstrated leadership and fostered the development of a culture of curiosity by engaging with the organization in advance of the project. This approach not only encourage buy-in or commitment to transformation but generates additional creativity to the project. Four action items of this strategy follow.[68]

1.10.3.1 Before Launching an AI Project

Throughout this book, we exam case studies, lessons learned, and other good/best practices available to all. Sometimes, readers will note there is some repetition. The fact that several or many have followed similar procedures suggest that they do work and are applicable in a number of environments. Coca-Cola's asked and answered several key questions before undertaking it GenAI project:

- **What Capabilities Does the Organization Need to Build/Acquire First?**— Projects should not start from ground zero. Certain activities, training, team building, and assessment should take place during Phase O. The ground must be laid before an organization can 'hit it running.'
- **Where Is the Line Between Human Creative Processes and Those Relegated to AI?**—There is a balance to be struck. It must be identified

and communicated to humans as appropriate and timely. AI is not the be all, end all. Not yet anyway and the jury is still out as to if and when.

- **How Will We Address Legal, Ethical, and Even Moral Questions?**— These are all soft issues as opposed to technical or engineering. Major areas include Intellectual Property, Data Use and Ownership, and Contributor rights. Clarity and plans matter set expectations and protect the corporate and product/solution brands.
- **Are AI Solutions Aligned with How People Emotionally Connect to Our Brand?**—Perhaps for this company, they have taken the lessons to heart of the New Coke product roll out of 1985. Brand visibility and loyalty are expensive; time consuming to build and can be lost in an instant. Technology should not be a casual event of brand damage or even destruction.

It is important that leadership trust the organization and key suppliers and customers. In a social media age where misinformation or incorrect understanding can overwhelm an initiative or project almost instantaneously, leadership, not simple management, is THE key to success.[69]

"One of the things we keep repeating in Coca-Cola is it's about AI and HI. It's *artificial intelligence* and *human intelligence* and *ingenuity*. I really believe our opportunity as humans is to continue to work on the creative side, on the values side, and use AI to scale ideas."[70]

1.10.4 AVIATION SAFETY

We all fly, and many of us still have a fear of flying. The United States' National Transportation Safety Board (NTSB) reports that there have been 20,684 incidents between the 15-year period from 2008 to 2023. Of these, 3,764 were fatal accidents with 6,254 fatalities. There were also 3,745 injuries.[71] According to the NTSB Monthly Aviation Dashboard, through April 9, there have been 153 aviation accidents with 23 deaths during the first quarter of 2025.[72]

Originally created for the healthcare sector, one Israeli firm's technology has become a game-changer in Energy, Transportation, and Industry 4.0. The value proposition—enhancing safety, predicting hazards early, and reducing maintenance costs. According to their CEO, "With advanced micro-cameras and specialized AI models, we provide predictive maintenance and monitoring in the most challenging environments." Condition-Based AI-Powered Vision-Based Predictive Maintenance systems acquire and analyze huge volumes and categories of real-time telemetry. "As the aviation industry moves towards intelligent infrastructure and advanced AI technologies, the integration of predictive maintenance solutions could potentially revolutionize the safety landscape."[73]

AI is poised to play a major role in general and commercial aviation, from the passenger, operations, flight, and maintenance. GenAI implementation required management to meet these five challenges. All are applicable to safe flight operations.

- **Data Security and Privacy**—Data is foundational and in addition to high quality, as core to the organization, it must be secured and privacy regulations and ethics forefront.

- **Accuracy and Reliability**—Basically, data validity and reliability are the core quality requirements for all data sets. This includes addressing bias and misinformation problems.
- **Integration Complexity**—New AI systems must work with existing legacy systems. This is not always easy, and integration strategies must be well thought-out and executed.
- **Regulation and Ethics**—AI regulations are lagging behind this technology explosion much as it is with all new highly disruptive technologies. Likewise, ethical questions must be agreed to. Organizational governance models need the flexibility to incorporate new thinking without losing their core organizational mission.
- **Cultural Impact**—Humans are part of the AI critical path and robust change management is the key to making AI-driven organizations a true reality.

All of these points are applicable in other sectors as well.[74] Each and more will be addressed throughout this book and expanded accordingly.

The CEO of United Airlines believes the company is "probably doing more AI than anyone." While much of their efforts are focused on customers and labor contracts and seemingly going well. They admit to some missteps implement AI-enabled preventative maintenance. The airline is continuing to gain traction with this business and technical work process application.[75]

The business results from investment in AI and the scale to the enterprise level remains elusive. Not all processes are good candidates. The quandary of 'go-with-your-gut' versus solid capital economic assessment is active in AI right now and possibly to the detriment of some organizations. We address this further in Section 4.2.5.2 'AL Value: Enable Measurable Business Gains' of Chapter 4. Safety and maintenance often fall in this category with over but more frequently under investment not just in funding but in managerial focus.

This case is an excellent example of best practices cross over from one sector to another. One suspects that in three years, AI best practices knowledge available to all from most major sectors will be encyclopedic.

1.10.5 TRENDS IN HEALTHCARE

During the writing of this book, I went to my physician for a routine follow-up. He was using a new AI solution, where the prompts appeared to be my answers to a series of questions, not all of which only needed simple responses. The doctor confessed that he was still learning how to use this new technology, and he seemed genuinely excited about it.

One thing struck me, which we have discussed herein. He was one of only a few that were using the technology. Effectively a pilot that enabled the hospital to learn about the benefits and limitations of the AI solution in a controlled environment. Lessons for all us as we roll out new AI solutions.

Our health is an area where AI will most likely touch everyone. It is also an area where accuracy is critical. Like most applications, AI is rapidly growing in this highly regulated sector. This writer has limited experience with medical devices and understands that the Food and Drug Administration (FDA) approval for use with humans can be a long and arduous journey. The US FDA has expresses a lot of interest in AI and ML for use in healthcare. This is a complex regulatory environment beyond the scope of this book. Interested readers may want to investigate Software as a Medical Device (SaMD) as it relates to AI and your local regulations.[76] Also see Section 4.2.1 'Medical Superintelligence 101' of Chapter 4.

In this section, we address a few of the notable use cases available as this manuscript is being drafted. Predicting that a large number and perhaps more earth-shattering discoveries for AI will be found, perhaps even by readers of this book.

1.10.5.1 AI Healthcare Benchmarking

There are efforts underway such as OpenAI's HealthBench "to better measure capabilities of AI systems for health. Built in partnership with 262 physicians who have practiced in 60 countries, HealthBench includes 5,000 realistic health conversations, each with a custom physician-created rubric to grade model responses."[77]

Look for others to emulate and we expect other sectors to continue to build effective benchmarks. Hopefully, this will enhance acceptance and trust by practitioners. Benchmarking should be a component of the CAIO's arsenal to assure AI is successful within the organization.

Similar constructs can be developed for all 16 Critical Infrastructure sectors and others as well. The more complex, the more useful AI/ML tools will be.

1.10.6   CONSULTING AND PROFESSIONAL SERVICES

"Life can only be understood backwards; but it must be lived forwards."—Søren Kierkegaard (1813–1855).[78]

A deceased close friend, once an employee of several major software and professional services firms once bemoaned a decline in business. He asserted that the consultants had trained their clients in certain areas and thus were no longer needed. Is AI training end users and having the same effect on professional services?

Over a quarter of a century ago, we launched a consulting business. Our focus remains on operational excellence primarily for heavy industry and those whose product lines and services add value in Homeland Security's 16 Critical Infrastructure sectors.[79] Like most firms, we quickly developed tools that enabled our delivery as well as differentiators from the raft of so-called consultants—every retired or unemployed executive/technologist or engineer. There has long been hypercompetition in this space. We quickly learned that software was a significant barrier for our competitors.

Once a process becomes repeatable and the only thing that materially changes is the data, software is the best logical solution to the problem.

Moreover, we worked closely with clients. This assured that the solutions developed solved real-world problems and were not the dream of detached SMEs. To date, we have four major Cloud solutions: Cross-Cultural Training, Economic Value Assessment, Process Simulation, and Operations Management.[80] Other solutions include Maturity Assessment, Systemic Safety Index, and other more focused explications.[81]

This business model is not unique to our firm. Most others have adopted similar models including all of the major large-scale management consulting firms. Many others have their roots in accounting and decades ago developed expertise in financial management as well as enterprise resource planning (ERP) software products.

1.10.6.1 Enter DOGE

The United States Congress has a long history of putting bills of thousands of pages out for vote the next day. A long-standing process designed to force agendas down the throats of their constituents. So it was in December 2024, when a 1,500 plus page Continuing Resolution supposed to stave off a federal government shutdown demanded immediate passage, or else!

Led by Elon Musk, the Department of Government Efficiency (DOGE) responded, and a much shorter (100+ page) bill or legislative proposal was passed. How was this possible when so many other gambit met with congressional acquiescence? Could it be that AI had a role in this Creative Destruction?[82]

The 'Jig-is-Up,' going forward using sheer paper mass will no long be the driver in the lawmaker approval process. The "We have to pass the bill so you can find out what's in it" era is over.[83] Bills of almost infinite length can be read in moments and even determine what part of a bill were written by other AI programs.

The DOGE federal government agency, by agency auditing, appears to be taking a similar approach. This ability enables the disassembly of huge multi-volume consulting and engineering proposals and project deliverables. We will most likely look back on this period as game-changing for even the largest clients and professional services firms.

By all accounts, DOGE is a relatively small group of individuals with strong AI skills.[84] If they can audit a 1,500-page bill this quickly, why can't others do the same? Good question, since it did not happen prior to December 2024 and there have been many massive bills put forth and passed into law, largely unread.

1.10.6.2 AI Impact

The operations of even the largest consulting and engineering firms are not very different from that described for small firms. Typically, a large number of (best and brightest) junior consultants (often titled analyst or associate) work under a partner and his or her management team to deliver client solutions. Much of this work is manual, time and materials billing for research, data gathering, and benchmarking. Repetitive and time consuming by nature.

One can expect in the near term that as with other similar repetitive tasks, GenAI, Big Data Analytics, and Automation will provide an 'assist' during the work processes with expectations of a better deliverable. Likewise, new entrants will be expected to have a level of AI proficiency that will differentiate them from otherwise equally qualified candidates. This move toward AI advisory work will become the foundation of Creative Destruction throughout the sector. Within a few years, much of the entry level consultant work will be automated, thus requiring fewer new consultants. Most likely the traditional pyramid of Partners, Mid-Level and Entry Consultants will flatten with smaller more specialized teams. Mid-level responsibilities and deliverables will increase as will the need for different skill sets such as engineers and data/AI specialist. These teams will deliver a higher end product, faster and with lower billing rates. A shift toward value pricing and outcome performance will replace the old time and material billing process clients have hated for years.[85]

In this author's opinion, this sector may revert asymptotically toward its historic roots. Smaller, more focused firms will provide counsel using deep expertise using AI to add significant value to clients, quickly with an increased likelihood of success. There will still be a role for Deming's, Drucker's, and other true oracles, but this expertise will remain available from individuals directly and not as part of a broader engagement.

The same will be true for engineering, legal, medical, therapy, and most other business/technical areas of expertise. Clients will purchase units of AI-driven expertise on demand and not as part of a large extensive effort by large teams. More like special operations as opposed to large scale maneuvering on the corporate battlefield.

For many years, software development has moved toward smaller multi-functional high-performance teams as opposed to large armies of programmers with a managerial hierarchy. AI will most likely accelerate this trend across all Knowledge Workers.

Perhaps we foretold this over three decades ago in our 1993 peer review publication, *Scientific Management and the Knowledge Worker* when we stated, "We live in a time of rapid technological change requiring the re-education of a large segment of the population. This is precisely the environment which fostered the development of Scientific Management."[86] We will discuss Scientific Management in more detail through the lens of its actionable cornerstone, 'Scientific Method for AI' in Section 3.1 of Chapter 3.

Expect the results to surpass current operating model performance.

1.10.7 ADDITIONAL USE CASES

There is an emerging body of (use case) work. Many have nuggets of information that is useful and can be used by organizations, including those in different sector. However, the flood of these materials suggests that not all meet quality standards. Therefore, care must be taken when assessing their applicable to your situation.

The following list of over 2,500 use cases was posted on LinkedIn during August 2025.[87] It lists use cases of 12 major players in the AI sector. While it is commercial in nature, detailed reviews of how AI is deployed across a large number of organizations in multiple sectors may be of interest. It is provided with the usual caveats regarding 'provided for education purposes only and not a recommendation by the author and/or publisher.'

- **Amazon**—Generative AI Customer Stories[88]
- **Capgemini**—Harnessing the value of generative AI: 2nd edition: Top use cases across sectors[89]
- **Deloitte**—The AI Dossier: A collection of our latest high-impact AI use cases by industry and type[90]
- **EY**—AI Use cases[91]

- **Google**—601 real-world gen AI use cases from the world's leading organizations[92]
- **IBM**—The most valuable AI use cases for business[93]
- **Intel**—Artificial Intelligence (AI) in Manufacturing[94]
- **McKinsey & Company**—Beyond the hype: Capturing the potential of AI and gen AI in tech, media, and telecom[95]
- **Microsoft**—AI-powered success—with more than 1,000 stories of customer transformation and innovation[96]
- **Oracle**—Generative AI and its use cases for enterprise applications[97]
- **PWC**—AI-driven revolution: How to drive your business success with strategic AI use cases[98]
- **SAP**—SAP Business AI use cases[99]

Likely, this list will continue to be updated along with additional cases from other organizations. Something readers may want to monitor and scrutinize accordingly.

1.11   AI CONTEXT FROM THE PERSPECTIVE OF HISTORY

In our 2025 book, *Navigating the Data Minefields: Management's Guide to Better Decision-Making*, we devote Appendix A—*A Brief History of Artificial Intelligence* to a detailed discussion about the roots and history of AI. We invite interested readers to review that Appendix in the context of both that book as well as this edition. A review of the theme of the earlier work follows:

> *Navigating the Data Minefields: Management's Guide to Better Decision-Making* provides executives and SMEs with a 'reasonable' set of (useful) tools they can adapt to their specific organization and operating environment, which now incorporates Artificial Intelligence (AI) environment and its suite of solutions such as Generative AI (Gen AI), Machine Learning, and Large Language Models (LLMs), et al. While complexity can never be taken out of an integrated system, decision-making can be facilitated using metrics that take into consideration the quality of the data used to make the decision, i.e., risk mitigation.[100]

Briefly, we have been interested in the *idea of manufacture intelligence* for centuries. Furthermore, "Perhaps the most lasting contribution of these early researchers was not scientific but semantic. Artificial intelligence (AI)—and all the different, sometimes conflicting, ideas that term conjures—was originally meant to entice those who might bring machine-based intelligence into being."[101]

Perhaps the most seminal work in the field of AI took place in the early to mid-20th century and efforts of the British mathematician, Alan Turing.[102] "In 1950, the mathematician Alan Turing put forth this question. Rather than attempt to answer it using conventional logic, he proposed a new disruptive model—the Imitation Game. He posited whether computers could one day have the cognitive capabilities of humans."[103]

While the full history of AI is perhaps hundreds of years or older, its modern incantation has celebrated in 75-year anniversary. Much has happened since Turing laid down the Thinking Gauntlet, but most believe that AI passed the Turing Test. AI has a rich history and has a solid foundation on the shoulders of giants.

Interestingly, both the software industry and the early work in Game Theory was developed during the Turing era as well. We also believe that Game Theory is a fundamental underpinning of the Turing Test.[104]

A crowning event to date,

John J. Hopfield and Geoffrey E. Hinton were awarded the 2024 Nobel Prize in Physics for developing machine learning technology using artificial neural networks. In Chemistry it was awarded to Demis Hassabis and John M. Jumper for developing an AI algorithm that solved the 50-year protein structure prediction challenge. This highlights AI's impact on science, medicine, and society; however, the winners acknowledge ethical aspects of AI that must be considered.[105]

The foundation for a great AI future is well laid.

1.11.1 WILL AI REPLACE ME?

The short answer is yes, it will replace the daily tasks some perform. However, as with preceding technologies, it can open up entire new industries. Look at any sector since the dawn of the industrial revolution in the 1800s and you can see this pattern, i.e., automotive, aviation, medical, energy, and so on. Leading and managing these new entities will require change and growth as well.

All technology disruptions generate concerns about the impact it will have on the individual. When computer-aided design (CAD) emerged and threatened the role and livelihood of draftspersons, similar alarms were raised. As with every new approach, there were positive and negative attributes. Negative impacts included a reduction in the brainstorming process, sketching. However, the benefits outweigh the detriments and today CAD is widely accepted as fundamental to the design process.[106] The case can be made that for CAD and other more mature information technologies, the marginal costs of adoption and use are near-zero.

We seem to have a creator gene somewhere in our genome. Leaving aside religions (current and past), humans continually seek a relationship with a Higher Power, and/or creating protagonist partners/robotic servants. In Greek mythology, the Titan Prometheus befriends humanity and is even credited by some as our creator. For these transgressions, Zeus condemned him.[107] Later, Frankenstein turned on his creator, wreaking havoc. This 1818 novel, *Frankenstein; or The Modern Prometheus* and its derivatives continue to play a major role in western culture.[108]

- **AI and Warfare**—The ancient Chinese strategist Sun Tzu is credited with the recognition, "All war is based on deception." The theory goes that if war is primarily delivering overwhelming violent lethality at the appropriate time and place, the definition of AI can be said to be, *the use of vast amounts of existing and new data to deceive and confuse the enemy.* Likewise, if deception is a primary concern, the definition of AI remains the same.[109]

This use of technology is akin to the cyber and deep fake used by adversarial governments. When used by political parties, scammers, or terrorist, its use is mostly for personal gain.

- **AI Companion**—A former colleague of mine, the neuroscientist, David Eagleman has a weekly podcast, Inner Cosmos.[110] In his March 31, 2025, EP98 "What's the future of AI relationships?," the question is raised—How many people are having relationships with artificial neural networks? The surprising answer is quite a few, hundreds of millions (perhaps a billion). Several key points were discussed:

  1. People care about other people as well as what they think of us, and our brains have developed over millions of years to interact and belong (form relationships) with others. Our neural circuitry 'wires us that way.'
  2. **R**elationships are **C**ondition dependent, i.e., spouse, manager, and friends each with specific sets of **B**ehaviors—**The RBC Model** (Section 2.3 of Chapter 2). This is individual dependent, thus all of us have thousands of relationship interactions on a continual basis.
  3. We only have one way of understanding these relationships or socialization and that is to model other people within the scope and limit of our own understanding of ourselves and biases.
  4. We love our technologies, i.e., automobile, smart device, boat, etc.—sometimes referred to as 'toys.' Our capacity for feelings for non-real humans is not new.
  5. Relationships with AI companions or replicas run the gamut of human behaviors, loneliness, mentor/tutor, friendship, romance, just sexual, confidant, etc. A mirroring effect or mental model as well.
  6. There is some evidence that AI companions can have a positive effect preventing suicides.
  7. However, our brains do not respond exactly as they would to the living flesh and blood human. It appears total dependency on AI companions still leave a gap that needs fulfilment.
  8. AI companions may serve some value as so-called sounding boards that may even 'push back' with Tough Love when they deem it necessary. Perhaps a parental figure? The extreme may even be an abusive relationship with an AI Companion, or the companion as the abused.
  9. One analogy may be the role pornography plays in society. Perhaps a degree of depravity but maybe filling a valid human need? Readers can make their own judgment of the controversial subject. It is offered for information purposes only.[111]

If this is a subject of interest, readers should review the large body of work from David Eagleman and others seeking to develop deep insight into how humans will relate to AI. Perhaps AI will integrate into human psychology more deeply than any other human-created object.

All of these points will eventually be tested at the society level of scale. One suspects significant controversy regarding the role of AI as companions. When one develops a large spreadsheet financial model or authors a book, there is a certain amount of Pride-of-Ownership. Most likely similar relationships will develop for developers and users of AI apps. Gamers (users) are very devoted to their software products and their position in gamification hierarchies.

As of this writing, there are over 100 AI applications offering romantic and sexual companionship. Like all AI apps, they learn from continuing contact with their human users. Even the appearance of love. However, there are concerns regarding emotional dependencies as well as hindering human-to-human social skills, including emotions.[112]

There are two opposing explanatory hypotheses regarding the effect of online behaviors on adolescents' well-being,

> The **displacement hypothesis** predicts that online communication reduces adolescents' well-being because it displaces time spent with existing friends, thereby reducing the quality of these friendships. In contrast, the **stimulation hypothesis** states that online communication stimulates well-being via its positive effect on time spent with existing friends and the quality of these friendships.[113]

For this one 2007 study, support was found for the stimulation hypothesis but not for displacement.

In many ways, these illusions of understanding are similar to gaming scenarios. These are the reasons games can be so intriguing and all-consuming as well as excellent training or serious education instruments.

In one case, a subject of my Game Theory (Section 6.6 'Case Five—AI-Driven Cross-Cultural Serious Gaming' of Chapter 6) research became visibly upset when the game did not develop according to his expectations. This incident confirmed to his writer that humans get into character and the scenarios in which they are involved. AI should be no different.

It is fair to surmise that humans can and will develop relationships with AI apps. The real test is how society and by extension organizations deal with this phenomenon. Is this a form of addiction, something else, or natural? How one views this will dictate his or her response. An organization's perspective will most likely have a significant impact on its culture.

- **Tasks' Automation**—According to the consulting firm McKinsey, by 2030, approximately 30% of the tasks performed in the US economy can be automated. GenAI is accelerating this trend with the biggest hit coming to office support, customer service, and food service sectors.[114] This suggests that during the remainder of this decade, individuals will need to transition to those sectors exhibiting growth or at least stability.

  Moreover, AI solutions will likely impact each sector inconsistently, meaning that its impact will vary and in the case of declining fields may likely speed those adverse impacts. This may place more emphasis on the use of AI in these lower margin sectors and require a workforce that is very conversant with AI.

  Similarly, growth sectors may see AI accelerate their surge. Individuals with AI expertise may see their careers rise faster than others. One example,

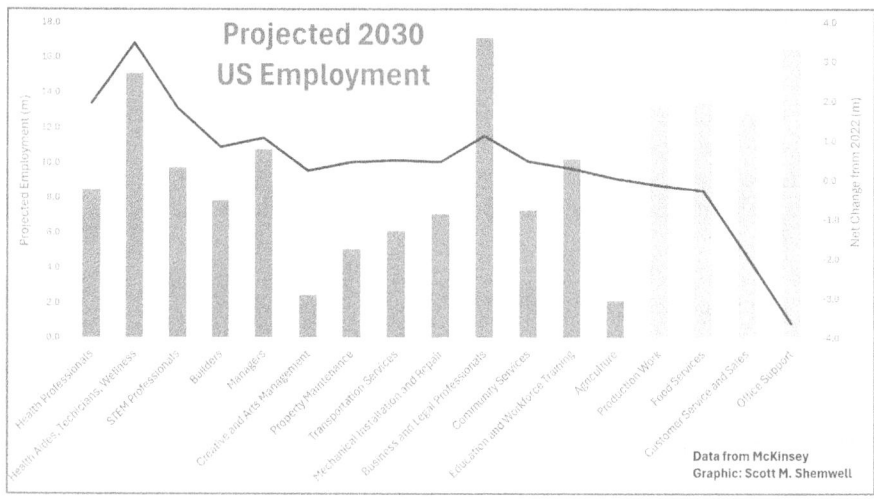

FIGURE 1.4   Projected 2030 US employment.

in the early 1980s, my relatively limited business expertise in western Europe gave me a leg up when the need for sales expertise in Asia was required. This promotion was career changing for me as I excelled in Asia. All of us need to stay relevant throughout our careers. Look for opportunities when they least likely appear.

McKinsey has some nuances with their growth data but for our purposes, this simple employment chart is satisfactory.[115] Note that the chart bars are a different color for those sectors with either zero or negative growth expectations (Figure 1.4).

The takeaway—AI is playing a major role in all economic sectors and employees should prepare for sometimes dramatic changes in their workplace. Employers will need to put training and other processes in place to improve or upskill the existing staff expertise as well as assure future prospects find their place of employment an exciting and career fulfilling environment.

I always hated Spellcheck and despise AutoComplete/Correct. I quickly found myself relying on the software to fix my typographical and outright spelling errors and still do—a marriage of convenience. On the other hand, AutoComplete finish statements that are often at odds with my intent, and I have found numerous mistakes from this so-called time saving tool requiring more time to find and fix than might be saved.

AI exhibits even more potential value and convenience as well as greater foibles. "In our 2023 book, Smart Manufacturing: Integrating Transformational Technologies for Competitiveness and Sustainability, we asked the question regarding Machine Learning. 'What do you do if your machine is learning the wrong things?'"[116] We addressed this issue from three perspectives in our 2025 book, *Navigating the Data Minefields: Management's Guide to Better Decision-Making.*[117]

- **High Reliability Management**—Organizations with robust systems that consistently enable the accomplishment of critical goals while avoiding yet rapidly responding to errors that might cause catastrophic events (Section 6.6.1.2.3 'Outcomes of the Game' of Chapter 6).[118]
- **Error Management**—The late James Reason's *12 Principles of Error Management* apply to AI technology development.
- **Contingency Planning**—Response scenarios in advanced for AI caused potential incidents, i.e., Three Mile Island, Fukushima, Deepwater Horizon, etc.

These are important processes for business management as well as AI development management to undertake, rehearse, and refresh. As Louis Pasteur is credited, "Chance favors the prepared mind."[119] This quote encapsulates the three processes. Organizational governance and risk mitigation policies and procedures is a great place to start this effort.

This is a best practices approach toward making sure organizational and human brains do not atrophy as is often the case when technologies become taken for granted. Typical is a 2025 Wall Street Journal article states,

> Like AI itself, research into its cognitive effects is in its infancy, but early results are inauspicious. A study published in January in the journal Societies found that frequent use of AI tools such as ChatGPT correlated with reduced critical thinking, particularly among younger users. In a new survey of knowledge workers, Microsoft researchers found that those with more confidence in generative AI engaged in less critical thinking when using it.[120]

Is this different than using the MS Excel spreadsheet? Probably not as much as one might think. The problem statement is still at the crux of AI, at least for now.

Questions that must be asked and are often given short shrift:

- What happens if AI gives the wrong answer and how would one know?
- When does the human intercede?
- What if the human intercedes and is wrong?

Both management and employees/contractors must have predetermined answers to these questions, especially when AI is used to manage critical infrastructure systems. Hopefully, the answers to these and other AI conundrums will become clearer in this book.

Finally, an argument can be made that AI is ingrained into the human psyche. Paraphrasing Voltaire, "If AI did not exist, it would be necessary to invent it."[121]

So, will AI replace me? Probably not, although there is little historical evidence that job descriptions and even lifestyles do not change given the advent of overwhelming technologies, i.e., Smart Devices, automobiles, aircraft, etc.

The likely scenario is this extension of the human will usher in new eras of dramatic human advancement. Each individual has choices to make as to how to take advantage of what will eventually be a high-value add to life at near-zero marginal cost. This promise is not just game-changing, it is humanity-changing, akin to the use of fire, development of the wheel, and other technologies at those fundamental levels. Perhaps raising many to Maslow's level of Self Actualization, and everyone above poverty levels (Section 2.5 'Maslow's Hierarchy of Needs' of Chapter 2).

Additional details on these subjects are available in our 2025 book, *Navigating the Data Minefields: Management's Guide to Better Decision-Making.*[122]

1.11.1.1 AI-Powered Engineers

Agentic AI, aka autonomous AI, can run independently to design, execute, and optimize workflows. In this capacity, it can make decisions and achieve predefined goals with little or no intervention.[123] Since they do not require prompts such as needed by GenAI systems, they are well suited for real time and other autonomous operations. At their core are five features: Autonomy, Goal-Oriented Behavior, Adaptability, Collaboration (with human and non-humans), and Environmental Awareness (using multimodal data). However, there are risks associated with autonomy and complexity in agentic AI algorithms.

- Potential misalignment with human and organizational values and their unintended consequences
- Loss of control that may result in unpredictable behaviors
- Ethics, biases, and other accountability/privacy concerns
- Security and/or malicious use
- Economic disruptions from errors, mistakes, or human job transition

Governance and strong software/project management processes are critical. Moreover, so as these systems will work more closely with operations and manufacturing where unplanned downtime and safety are major valid concerns.[124]

According to one oil and gas industry source, these so-called "AI-powered engineers" are addressing the industry's decades long huge data management dilemma. Up to 70% of an engineer's time is devoted to data management. Finding and normalizing data before engineering processes even begin.[125]

This requires both engineering and data domain expertise and is not easily outsourced either to the IT department or third parties. The data is usually very sensitive and part of the organization core knowledge base. AI-powered engineers are best practice solutions.

1.11.2 AI and Expert Systems: Early Myths, Legends, and Facts

It is often useful to review documents and opinion from a previous time. This can help readers understand the thinking of that era as well as how technology is framed in a sustainable way.

> The first AI programs were written in such a way as to arrive at the solution of a problem by 'reasoning' through a series of logical propositions. It was in the early 1980s that a different approach emerged: knowledge-based systems. Also called Expert Systems, these artificially reproduce the performance of an experienced person in a given field of knowledge or subject.[126]

The following is taken (direct quotations in *italics*) from a 1990 article published by the Institute of Electrical and Electronics Engineers (IEEE). In this article titled, AI and Expert System Myths, Legends, and Facts.[127] Mark Fox of Carnegie Mellon University identified three AI categories:

- **Myths**—*Perceptions not based on any fact*
- **Legends**—*Perceptions, once based on fact, that have been blown out of proportion*
- **Facts**—*Perceptions that have a real basis in fact*

The author stated, *"By identifying these perceptions and misperceptions, it is my hope that we can use AI technology selectively and successfully to increase the quality and productivity of decision making."* He categorizes the following.

1.11.2.1   Facts

- **Search Is a Core AI Concept**—A Physical Symbol System must have the necessary and sufficient means for general intelligent action. Evolving (iterative) symbol structures exercise its problem-solving intelligence by searching.
- **AI Reduces Search Combinatorics by Applying Situational Knowledge**—*Domain knowledge can be represented as elaborations of operator conditions (that is, descriptions of situation in which a particular decision is to be made), which makes the application of operators more sensitive to the current context.* In this context, situation is equivalent to Conditions in the RBC model (Section 2.5.1 'Categories' of Chapter 2).
- **AI Reduces Search Combinatorics by Reformulating Problems**—Expert systems work well when problems are small and stable. More sophisticated problems require more in-depth search capabilities. By reformulating the problem as a simpler task, this solution becomes the guide to solve the original problem statement.
- **AI Enhances Search through the Use of Opportunism**—In resource constrained situation, optimizing resource allocation is more effective than scheduling time constrained (calculation) jobs, perhaps analogous to Pareto Efficiency (Section 3.1.1 'Human Input Still Needed' of Chapter 3).
- **Knowledge Representation Is a Core AI Concept**—AI's ability to solve problems is derived from its ability to search for patterns. Pattern recognition is focus for knowledge representation.
- **AI Knowledge Representation Extends Quantitative Models by Abstraction and Differentiation**—These qualitative abstractions of underlying quantitative model answer question the underling quantitative model may not. See 'Latent Construct,' Section 1.12.2 for more detail.
- **AI Knowledge Engineering Tools Increase Productivity, Thereby Reducing the Cost of Creating and Maintaining Software**—AI programming increases productivity using higher level data structures, higher level languages, fewer interface modules, and reducing programing development time.

1.11.2.2 Myths

- **Expert Systems Differ from AI-Based Systems**—Fox defines an expert system as, "A computer program that emulates the search behavior of human experts in solving a problem. The important point is that experts solve problems by searching a problem space. What distinguishes expert search behavior from naïve search behavior is the rich set of operators from which experts choose when solving a problem."

- **If We Have an Expert, Then We Can Create an Expert System**—In addition to subject matter experts (SMEs) that understands and defines the problem, system software applications must include well-defined inputs addressing defined management problems, outputs well thought-out so that the application provides relevant information/knowledge.

- **All Expert Systems Are Expert Systems**—Often an interim expert output is the input to more traditional problem-solving solutions. In this sense, it is not a pure expert system but a hybrid.

- **Expert Systems Do Not Make Mistakes**—Expert systems emulate the human expert problem-solving process and thought structure. Systems take steps to minimize these occurrences by learning, i.e., ML. Some algorithms infer the conditions of rules from data (remember this article was written in 1990).

- **AI Replaces Conventional Approaches**—Some optimization algorithms such as linear programming provide better answers than AI 'satisficing' method.

- **Small Prototypes Can Be Scaled Up into Full-Scale Solutions**—Early prototypes can make good demonstrations; however, scaling can be an issue. For example, "When knowledge is considerable and highly interdependent, classificatory problem-solving methods do not scale up." Furthermore, In the case of internal medicine diagnosis tools, they may use a different algorithm when handing complexities versus the simpler prototype demonstrations.

- **Managing AI Systems Differs from Conventional Project Management**—The risk-driven Spiral Model appears to be the most suitable software development methodology for conventional and AI systems alike. While complex, AI development is still a project requiring strong, knowledgeable project management.

- **All AI Tools Are the Same**—There are many AI tools available with different capabilities and approaches. Develops need to conduct robust Technology Assessment (Section 3.3.4 'Technology Readiness Assessment' of Chapter 3) to determine which is best for the problem being addressed.

- **Learning an AI Knowledge Engineering Tool Is All We Need to Know about AI**—AI solutions are composed of several or many subfields with many theories and techniques. Knowledge of AI Theory is necessary in addition to the mechanics of AI programming. This issue is similar to the current need for SMEs to assure programs actually solve the problem they are intended to (Section 3.3.2 'Problem Statement' of Chapter 3).

- ***AI Knowledge Engineering Tools Are Good for Only AI Applications—*** *AI software supports qualitative and quantitative reasoning equally well. The bottom line is that we can use AI programing environments to build any application.* This has significant value to the overall software development sector as AI dramatically ramps up (2025–2035).

1.11.2.3   Legends

- ***AI Systems Are Easy to Build—***To the extend the problem can be mapped into the software, solutions can be relatively easy to build. However, for design intensive or difficult problems, systems will be more complex, costly, and time consuming to develop.
- ***Rapid Prototyping Leads More Quickly to Final Solutions—****Rapid prototyping elicits the requirements and specification of software for ill-defined problems; its importance has not been lost on software engineering in general.*
- ***AI Systems Can Be Easily Verified and Validated—***Software Verification and Validation (IV&V) is still required by AI solutions. Repeated testing over many problems will either verify and validate the software product or not.
- ***AI Systems Are Easy to Maintain—***Rule-based systems require that various rules depend/interact with other rules. This interdependence makes rule-based systems more difficult to maintain.

Fox states that he drafted this article referring to AI and Expert Systems,

> They are not simply laboratory curiosities. Nor are they panaceas for the many problems we face today. Instead, they are viable technologies that provide a fresh approach to solving many decision problems. Only by removing myths, laying open legends, and recognizing facts can we develop the technical and managerial skills required to apply AI successfully.

It is interesting to compare this line of thinking from the 1980s to our AI belief system today. We have a lot in common with our ancestors. Wise AI management and technical executives, users, developers, recommenders, and others will take these lessons to heat and learn from them.

1.11.3   CROSS-CULTURAL LEADERSHIP

Our multi-cultural world has always presented challenges when two or more individuals/teams/organizations engage with one another either in conflict or more peaceful endeavors. Driving daily behaviors much less transformational behaviors can be difficult and frustrating both for the change agent or manager as well as the individuals tasked with performing various tasks.

Currently, many groups are self-segregating and increasingly seeking semi- or even full autonomy. Leading such groups which openly have their own agenda can be a Herculean task.

We even ask the question, does or will AI have a culture? If so, what will its collaborative role be with human cultures? Moreover, how will leaders extract high-performance from these machines? AI machines are designed to continuously learn and that is a great trait for a follower. Hopefully, biases (Section 1.5.2.1 'Fourteen Biases and Stereotypes') will more limited than human counterparts which requires less coddling.

We do not believe that many of the leadership traits and knowledge from an extensive body of work will need to be rewritten. Changed to keep up with the times and cultures, pop and otherwise but foundationally human nature has not changed much over many millennium.

It is also a high probability that AI cultures, if they exist, will have some of the same traits as their creators. Even AI generations from now, some of these residual traits will be found in the cultural DNA, just like in their human counterparts.

In our June 20, 2025 blog, *Leadership: Pattonesque Style*, we make this point regarding this World War II icon.

> Old blood and guts, General George S. Patton Jr. gets a bad rap in my opinion. He probably was a pain in everyone's you know what, but he was paid for being something else. A leader who made consistently effective decisions of consequent, rapidly with decisiveness.[128]

However, as part of the Allies leadership team, he was not the most beloved and often at odds with others including his boss, General Eisenhower. Taking the mantle of leadership carries risk or criticism.

The effectiveness of a *Competent* and *Confident* leader is measured in the results of her decisions. In the high-performance, high-speed, high-risk, and high-consequence AI environment, aren't these the leadership traits we want rather than the typical more bureaucratic approach often taken?

In Section 6.6 'Case Five—AI-Driven Cross-Cultural Serious Gaming' of Chapter 6, we describe our long-standing cross-cultural serious training game that is in the process of transition from a two-human party negotiation to one where the human team works with an AI agent using role-based prompts as the other party. We have found this training process to be very effective as it is a realistic simulation of multi-cultural interactions to solve difficult global challenges.

The following two subsections offer insight into how AI is being used to unveil aspects of ancient writing and cultures. We are of the opinion that such tools can add value in contemporary attempts to better understand one another.

1.11.3.1   Enoch

> Writing is more than just a tool for communication; it is a reflection of our culture and identity. Cultural factors play a significant role in shaping writing practices worldwide, influencing language diversity, setting norms, and establishing rich writing traditions… Language is the vessel through which culture is conveyed, and it profoundly impacts writing practices.[129]

Rightfully, humanity has a strong interest in our roots. We seek our genealogical ancestral footprints as well as our search for why we are here. As I understand it, all major religions, philosophies, and belief systems are grounded in history as described by ancient scribes, stone masons, architects, philosophers, scientists, engineers, and others across the whole range of human expertise.

Moreover, cultures can trace themselves by counting millenniums as opposed to years, decades, or even centuries. As expected, artifacts from early eras are often clouded in mystery, seemingly mutually exclusive events and often leaves us with gaps and sometimes even more questions than when we started.

Language is complex. Research shows that there are three major linked/inter-related theories covered by the Integrational Theory of Language (ITL), i.e., the framework of a chain of language systems.

- **Integrational Theory of Linguistic Systems**—System aspects of languages
- **Integrational Theory of Linguistic Variability**—Variety structure of language
- **Integrational Theory of Language Use**—Communications including semantics

Not surprisingly, the Integrational Theory of Language Use is the least complete. One glaring weakness is in the area of 'inter-personal discourse.' To date, the emphasis of ITL has been the spoken language with the work on written language lagging.[130]

In an era when machines listen and voice to text is common, is this a weakness that can negatively impact the validity and reliability of machines conducting this transformation? We will discuss in more detail the impact on AI that diverse languages may have. Another possible source of bias and other errors.

Multi-disciplinary science has always played a major role in our search for the beginning. Now, AI is a new and vetted tool to add to the archaeological quiver.

Using AI to investigate the Dead Sea Scrolls has open new knowledge about their age, possible authors, and timeline.

> "Pioneered by the University of Groningen, this multidisciplinary work merges radio-carbon dating, ancient handwriting analysis, and machine learning. The outcome is Enoch, the first AI system that can derive probabilistic dates from the script of Hebrew and Aramaic manuscripts, providing estimates based solely on the script's stylistic elements."

Using the neural network, BiNet Enoch scanned and analyzed features of the scripts curvature of ink strokes and shapes of letters. These predictions were validated against radiocarbon data. Some dates were confirmed, and others had their dating revised. The overall value of having a more granular and accurate dating provides new insight into ancient dynasties and development of emerging religious movements.[131]

This technique may be the dawn of new AI-enabled approaches to the human language. It may find use when dealing with unstructured written text data, including kanji, hanja, chữ Hán, Arabic script, Cyrillic, and other languages (including extinct ones) over the long period of the evolution of that language. In other words, place language of all forms and their historical context such that its contemporary interpretation is aligned with the original intent.

For example, the English language is constantly changing and words and phrases once considered 'slang' make their ways into the formal lexicon. This suggests that the language of this writer's youth is a different one than our contemporary written and spoken world. The same is likely for the children of the current generation. Moreover, this example is just one language. Each is undergoing a similar transformation.

The Enoch project focuses on reading text. Linguistics AI solutions can look at situational features of any mode of realization (spoken, written, and signed), perhaps as a function the meaning of text (behavior of the author and consumer) and hence the relationship between these parties.

We develop the Relationships, Behaviors, Conditions model in Section 2.3 of Chapter 2. This model will further illuminate issues with linguistics, intent and content helping us better understand our past and global culture similarities and differences.

Readers may wonder why we devoted this much space to this issue. In addition, putting content in its proper historical context, AI and its future processes and products will be heavily dependent on natural languages. Processing (NLP) which is rule-based modeling of human language integrated with AI technologies. It seems to this writer that data scientists, AI developers, and others need to understand this source of bias and/or error. Compare languages, apples-to-apples!

1.11.3.2   Aeneas and Ithaca

Epigraphy, the study and interpretation of ancient inscription has a new AI friend. There are two solutions to the game-changing AI capabilities enabling us to better understand antiquity.

First, **Aeneas** is the first AI model to contextualize ancient inscriptions.

Aeneas is a multimodal generative neural network that takes an inscription's text and image as input. To train Aeneas, we curated a large and reliable dataset, drawing from decades of work by historians to create digital collections, especially the Epigraphic Database Roma (EDR), Epigraphic Database Heidelberg (EDH), and Epigraphic Database Clauss Slaby (EDCS-ELT).

We cleaned, harmonized, and linked these records into a single machine-actionable dataset that we refer to as the Latin Epigraphic Dataset (LED), comprising over 176,000 Latin inscriptions from across the ancient Roman world.

This model use transformer-based decoder and specialized networks to handle the input of inscriptions and restoration and dating of the text."[132]

Likewise, "**Ithaca**, the first Deep Neural Network for the textual restoration, geographical and chronological attribution of ancient Greek inscriptions. Ithaca is designed to assist and expand the historian's workflow: its architecture focuses on collaboration, decision support, and interpretability." This AI solution can attribute inscription to the original location with over an accuracy of over 70% within a 30-year window of its writing.[133]

A modern-day Rosetta Stones, the keystone for decoding Egyptian hieroglyphics, Aeneas and Ithaca can be adopted to other early languages and inscriptions of all types from papyri to coinage. Likely, they should have more contemporary uses as well, perhaps in a *cross-cultural* setting.

Look for additional use of AI in these fields with exciting new understanding of our past and expanded biblical knowledge. The world's other major religions and philosophies should benefit as well.

1.11.4 Current Status and Future Trends

We can expect AI development to move toward a more human experience in that the hard technology is more mature than the empathy and thought component. As with other new technological marvels, the current situation and near-term future is somewhat cloudy.

This section could be an entire book in itself. A few examples of areas to be determined include the following:

- **Vibe Coding**—"A term coined by OpenAI cofounder Andrej Karpathy, is essentially when developers write code based on how something feels—with the help of AI. It's fast, flexible, and built on instinct. You're not locked into step-by-step rules. You move quickly and trust your flow."

  As *MIT Technology Review* explained: "Not all AI-assisted coding is vibe coding. To truly vibe-code, you have to be prepared to let the AI fully take control and refrain from checking and directly tweaking the code it generates as you go along—surrendering to the vibes."[134]
- **AI Fabric Architecture**—Offers has a lot of promise since it is modular and scalable. Its future proof capabilities offer a lot of promise for sustained operational excellence. "AI fabric is an evolution of the data fabric architectural approach. Data fabrics, which emerged in the past decade or so, combine the best aspects of data warehouses and data lakes. They provide a seamless, unified layer over an organization's data estate that makes data management quicker, easier, and more scalable."[135]
- **Data Centers**—A number of very high-profile investments in data center, some announced in the White House suggests that massive investments will be made. However, some are voicing concern that we may be in a 'data center bubble.' One pundit argues, "AI datacenters to be built in 2025 will suffer $40 billion of annual depreciation, while generating somewhere between $15 and $20 billion of revenue."[136]
- **IBM Research**—Findings indicate that organizations with a strategic approach to AI ethics can achieve an average ROI of approximately 13% compared to 5.9% for those without a cohesive strategy (AICadium).
- **Deloitte Insights**—Their survey reveals that companies with mature AI implementations report an average ROI of 4.3%, while beginners see about 0.2%. Leaders also experience shorter payback periods, averaging 1.2 years (Deloitte).

Moreover, ISO has a number of standards that help mitigate the risks and maximize the rewards of AI, including ISO/IEC 22989, which establishes terminology for AI and describes concepts in the field of AI; ISO/IEC 23053, which establishes an AI and ML framework for describing a generic AI system using ML technology;

and ISO/IEC 23894, which provides guidance on AI-related risk management for organizations. The regulatory framework is changing, and this is addressed further throughout this book.

1.11.4.1 GenAI and the Environment

AI is power hungry. Its explosive growth is demanding more energy. This section is numerical and data intense and was quoted directly from International Energy Agency report, Energy and AI. We did not want to lose the fidelity of the data by summarizing the materials. Its point to be made is that AI consumers' massive amounts of electricity, water, and other power generation support. This is why there is so much media coverage about massive data center capital investments.

For example, when training AI,

> estimates put the training data for GPT-4 at around 4.9 trillion data points, and the training compute at around 22 trillion calculations (that is, 2.2e25).
>
> Training is a time-consuming and energy-intensive process. Training calculations are performed on specialized computer chips such as GPUs. A single GPU can have a maximum rated power consumption of 1,000 watts in the case of the latest and most powerful chip. This is about as much as the power draw of a toaster. Large, state-of-the-art models are trained on clusters of many GPUs. For example, GPT-4 was trained on 25,000 GPUs with a combined rated power of around 10 MW. Additional power demand comes from information technology (IT) equipment operating alongside the GPUs in the servers used to train these models, such as CPUs, memory, networking equipment, and switches.
>
> Adding the power demand of additional IT equipment and the cooling equipment gives a total rated power of the equipment used to train GPT-4 of around 22 MW. This is equivalent to the power draw of around 150 high-power electric vehicle charging stations. It is estimated that GPT-4 was trained for around 14 weeks. Taking a load factor of 84% (Shehabi, et al., 2024), this results in a training energy demand of around 42.4 gigawatt hours (GWh), or around 0.43 GWh per day of training. This is equivalent to the daily electricity consumption of around 28,500 households in advanced economies, or 70,500 households in emerging market and developing economies. After training, models may undergo a process of fine tuning, which is much less computationally intensive than training and therefore less energy intensive as well.[137]

This poses significant challenges to electrical grids and as noted earlier, perhaps this bubble will burst. This need for electricity is a gating issue for all major AI deployments.

1.11.4.2 First Impression—Leading with AI First

The first point of contact for customers, partner, vendors, and others sets the stage as it is the organization's 'first impression.' How many of us have eventually hung the phone up after hearing one more time, "listen carefully, our menu has changed" and then nothing. Better yet is the slow, non-responsive chatbot that asks the same question repeatedly before passing you off to another bot that starts the process again from the beginning. Makes you want to buy from that firm doesn't it?[138]

What if an AI agent could process 90% of your issues quickly and accurately without all the current hassles? Such agents could also detect fraud, make decisions, and allow thousands of customers managed per employee. This new AI-First reality is here today and is rapidly becoming mainstream.

The technology is rapidly advancing and soon organizations can be 'AI-Native' as opposed to many of the so-called software Bolt-Ons designed to digitally enhance what we see today in the customer experience space. These AI agents will redesign entire processes, revolutionizing results.

By some metrics, management can expect up to 34X revenue per employee, a 90% reduction in customer acquisition costs, 16X increase in the product development cycle, and up to an 80% realization of non-financial benefits. Finally, AI-First industrial organization can expect greater than 15% improvement in working capital—enhanced liquidity.[139]

In the cited paper, BCG also identified many of the hurdles faced are similar to other AI challenges and include, Limited Awareness, Legacy System Limitations, No clear 'North Star,' Limited Expertise, Short-Term Conflicts, Governance Limitations, Fragmented Data, and Regulatory Hurdles. Against these challenges, there are dimensions for an AI-First organization that must be aligned with the overall business and technology framework. These include the following:

- **Governance**—Consistent with the approach defined here in Section 1.7 'Governing AI' and as applied to the organization.
- **Talent and Capabilities**—As always its first about the people not the technology. "People develop not only new skills, but also a new mindset."
- **Structures and Roles**—Both the human and AI roles as well as structuring the organization as a cross cultural learning entity.
- **AI Culture**—As defined in Section 1.11.3 'Cross-Cultural Leadership.'
- **Technology, Data, and Algorithm Foundation**—Again consistent with the challenges and concerns in Section 1.2 'Types of Artificial Intelligence Solutions.'

As with other AI implementation and deployment processes, AI-First must align and support the **mission** of the organization and add measurable **value**. These two metrics will require continuous monitoring and updating. Better customer relations will be worth this continuous improvement.

1.11.4.3 Model Merge

As discussed, the AI sector is relatively immature, and consolidation is expected. Generally, this discussion refers to suppliers; however, the number of models and overlapping technologies and products are enormous and still growing. Consolidation at these levels is already beginning.

Large language models (LLMs) have become increasingly capable, but their development often requires substantial computational resources. Although model merging has emerged as a cost-effective promising approach for creating new models by combining existing ones, it currently relies on human intuition and domain knowledge, limiting its potential.[140]

There are efforts underway make LLMs take better advantages of existing and future resources by shifting multiple paradigms into a single architecture. This model does not require additional training, which is a significant cost advantage. As might be expected, this has fuel interests and the Open LLM Leaderboard is dominated by merged models representing almost 4,600 models.[141]

The two key important points: LLM consolidation is a cost-effective way to continue development of new more capable models, and there are constraints caused by human limitations. This is an exciting approach that promises to reduce cost, complexity, and make LLMs more attractive to a larger business and research audience.

1.11.4.4   Data Center Power Requirements

AI currently consumes massive amounts of electrical power. Therefore, the demand for power is huge and growing. However, information technology has a historic curve. The marginal cost of computing has fallen for decades and along with that hardware technologies are more efficient and require less power per marginal compute unit.

> Will hyperscalers continue shouldering the cost burden, or will enterprises, governments, and financial institutions step in with new financing models? Will demand for data centers rise amid a continued surge in AI usage, or will it fall as technological advances make AI less compute-heavy?

The McKinsey research referred to has unknowns that cannot yet be quantified. Perhaps the AI-driven forecast models developed herein can be modified can help clarify future demand ranges with the associated risks profile (Chapter 4 'Risk Mitigation and Enterprise Alignment').

1.11.4.4.1   Driverless Vehicles

Whether one uses a so-called driverless vehicle or not, this use of AI will affect all of us in our everyday life. Versions of these systems are on the road today, even in large trucks.

Most likely, legal, ethical, and regulatory issues will continue to evolve with Advanced Driver Assistance Systems (ADAS). Proponents argue that a significant positive impact on safety is a major benefit as approximately 94% of all traffic accidents are caused by human error.

> The opportunity to reduce car accidents is making ADAS even more critical. Automatic emergency braking, pedestrian detection, surround view, parking assist, driver drowsiness detection, and gaze detection are among the many ADAS applications that assist drivers with safety-critical functionality to reduce car accidents and save lives.[142]

All late model automobiles and light trucks have some level of ADAS already. Figure 1.5 was adapted from the cited reference and puts this technology into perspective.

In my opinion, Conditional Automation (3rd Level) will likely be the minimum standard for new vehicles in the near future. It may even be mandated by law similar to the Corporate Average Fuel Economy (CAFE) standards.[143]

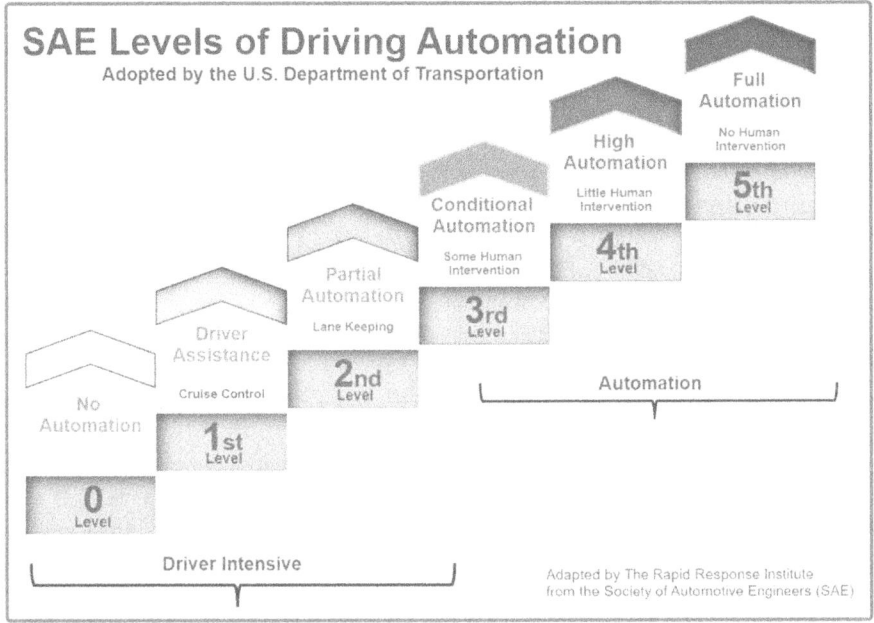

FIGURE 1.5 Automation systems for on-road vehicles.

1.11.4.4.2 Nuclear Revival?

The massive requirement for electricity has rekindled the nuclear power discussion. In addition to AI, electricity is rapidly becoming the preferred energy source as fossil fuels fall into disfavor. Moreover, public sentiment toward nuclear has improved since the high visibility incidents, Three Mile Island (1979) and Chernobyl (1986). Other less dramatic incidents include the Fukushima accident (2011).

Furthermore, nuclear technology has advanced, made safer, and construction and operating costs decreased. Small modular reactors (SMRs) "are a fraction of the size of a larger nuclear power plant, and they're comprised of standardized components that can be mass-produced in traditional facilities and shipped to the site,"[144] This is much better solution than the older large scale plant model.

It seems inevitable that nuclear will play a larger role going forward. AI should be a beneficiary of this solution that assures adequate electricity for our growing planet.

1.11.4.4.3 Prediction

As with any prediction, especially technology futures the vision is cloudy. The costs in time and money are very high. The stakes are soaring for all parties, power generations, transmission, and consumers. Building too much power generation leaves underused assets on balance sheets, while underinvestment may leave power grids in worse shape than they are today.

It is relatively easy to identify the current status of this technology. It is far more difficult to divine the future, even the next 18 months, much less than several years. Get our diving rods out of the closet.

1.11.4.5 Urania—AI Sensor Technology Breakthrough

In Greek Mythology, the Muse Urania, aka Ourania, one of the nine Muses and daughter of Zeus, is known as the Muse of Astronomy. The gods tasked her with the responsibility to decorate the night sky to inspire us mortals.[145] The more recent Max Planck Institute's AI algorithm named Urania is making significant leaps in the field of *gravitational wave detection*, first predicted by Albert Einstein's General Theory of Relativity, the understanding of how gravity affects the fabric of space-time.[146]

The core of this theory is a set of *ten differential equations*, the Einstein field equations (EFE). EFE describes how matter and energy determine space-time curvature and predict the existence of gravitation waves (ripples in space-time caused the acceleration of massive bodies), first directly detected in 2015—one hundred years after their prediction in 1915. Solving EFEs requires supercomputing efforts and Numerical Relativity Techniques. Importantly, EFE describes the large-scale structure and even the evolution of the entire universe.[147]

Urania is a major vault into an era where AI is not just a tool but a creative partner in scientific discovery. A few key points of value with benefits beyond not only this problem solution but the scientific community in general.

- Transcending traditional methods elevating the standards of sensitivity and detection ranges.
- Redefines the landscape of detector design, thus augmenting human ingenuity with the computation prowess to analyze vast multidimensional solution spaces.
- Incorporating new innovative scientific and engineering designs into existing detector array frameworks dramatically expands the boundary of human knowledge.
- This continuous optimization process has benefits across other disciplines using scientific, engineering, or economics optimization to address problems, i.e., economic utility theory.

> The influence of Urania extends beyond academia, fostering new opportunities for collaboration and innovation within the scientific community. By making 50 of its most successful designs publicly available through a comprehensive 'Detector Zoo,' the initiative encourages an open exchange of innovative ideas, attracting expertise from wide-ranging fields to enhance and refine these designs further. This approach exemplifies a commitment to open science, ensuring that advancements in technology and understanding are reachable by all, thereby expediting worldwide scientific progress. The practical application of these designs could lead to breakthroughs not only in understanding cosmic events but also in fields currently unimaginable.[148]

The implications for general sensor technology suggests that ML is a catalyst for future scientific discoveries across a broad range of fields. Machines will contribute novel solutions, while humans focus on understanding and integration into the existing set of frameworks. This is a significant step forward addressing and even solving complex scientific, engineering, and perhaps social problems.[149] Likely, steps forward with *interferometric* detectors that Einstein never contemplated.[150]

This is exciting because new research will make its way to our everyday lives in some capacity. Learning from this solution should help benefit all humanity.

Urania is demonstrating that AI solution will have a profound impact on complex problems from a disruption perspective and not only the evolution of technology.

Does AI Hold the Key to the Universe?

1.11.4.6 Choosing the Right AI Model

As with many subjects addressed in this book, it is a daunting challenge to address within the confines of a publication. We address the major type of models (Section 1.2 'Types of Artificial Intelligence Solutions'). Here we build on Section 1.2.1.6 'Model Selection Criteria' and although there is some overlap, this section expands with greater specificity.

It is useful for management to be able to ask the right questions of AI technologies when selecting the model to drive an application or even the enterprise. Key factors requiring careful consideration:[151]

- **What Is the Problem to Be Solved?**—This is the first and most important decision when selection the model type. The categorization of the problem dictates the type of model. For example, "if the data is labeled for input categorization, it falls under supervised learning. If the data is unlabeled and we aim to uncover patterns or structures, it belongs to unsupervised learning. On the other hand, if the goal is to optimize an objective function through interactions with an environment, it falls under reinforcement learning. If the model predicts numerical values, output categorization is a regression problem." Do not shortcut this exercise, it is critical to successful AI implementations.
- **How Will the Model Perform?**—Similar to above, performance is determined by the model selection. Metrics can include accuracy, precision, recall, etc. The available data sets can also impact on performance.
- **Can the Model Be Explained Easily?**—Depending on the scenario, a level of explainability is in order. A 'Black Box' mentality may not provide enough visibility into the solution.
- **How Complex Is the Model?**—AI models must be fit-for-purpose. The problems AI is asked to address can be complex (we describe several complex models in Chapter 6 'Capstone—Detailed AI Models under Development') However, complex models are more difficult to build, maintain, and cost more.
- **What Are the Data Requirements?**—Data sets are huge, and this is a massive issue. *Navigating the Data Minefields: Management's Guide to Better Decision-Making* should be consulted for more detail. Data SMEs must be part of any AI development team
- **What Is Feature Dimensionality?**—The vertical (volume) and horizontal (number of features in a dataset) need to be considered. Complexity is a function of data dimensionality and in some cases dimensionally reduction algorithms, such as principal component analysis (PCA) can be useful. PCA was established and I used it in my doctoral dissertation (behavioral analysis).

- **What Are the Training Requirements?**—Model training is critical and can be very costly. The question 'what is good enough?' needs to be answered. Like the statistical power curve, the last small percentage of gain can come at a great cost, i.e., is 97% accuracy at a cost of $10k ok or does the problem require 98% accurate at a cost of $100k? Also, don't forget to calculate the area under this curve. It is the quantitative metric for model performance.
- **What Is Performance of the Inference Engine?**—Inference speed refers to the time it takes a model to process data and answer the question. The application drives engine selection, i.e., a real-time decision can require instantaneous decisions (self-driving vehicle), while others may need greater 'thinking' (medical research).

Things AI developers must consider include, decision cycle time, hardware constraints, maintenance/updating, data privacy, robustness, and level of specialization of model, ethics and bias, model architecture requirements, and transfer learning (repurposing existing trained models).

Readers will note that this selection process requires a team of knowledgeable individuals to advise executives of recommendations. This process is both High Impact with associated AI Risk Governance Model (Section 3.2 'Operational Risk Management' of Chapter 3). Organizations should consider this CAPEX investment with the same rigor they require for all enterprise-wide decisions. This cannot be delegated to IT or the CAIO!

1.11.4.7 Cybersecurity

AI may be unique in that it can both help address cybersecurity issues as well as be the source of cyber vulnerabilities. CrowdStrike defines these AI-related components underpin current cybersecurity tools and strategies. The list is a direct quote and is in italics per citation style and convention.[152]

- *Machine learning (ML): To recognize patterns and learn from past incidents*
- *Natural Language Processing: To interpret human language, streamlining the analyst experience in task execution and democratizing security decision-making across teams*
- *Data Mining: To extract valuable patterns and insights from large datasets*
- *Predictive Analytics: To forecast potential threats based on historical data*
- *Behavioral Analytics: To monitor and analyze user behavior to detect anomalies*
- *Automated Decision-Making: To enable quick responses to identified threats*

Cybersecurity is a broad and rapidly changing environment. The human and AI team allows both do perform the duties they are best suited for; AI data volume and tedious work and the human is freed up to analysis that is outside the current capabilities of AI.

Remember that AI scouring the web for data may pick up contaminated data and even proprietary data. Appropriate data governance is also mandatory.

1.11.4.8   Fintech Open Source

Under the Fintech Open Source Foundation (FINOS), the critical banking sector has teamed with organizations like Amazon Web Services (AWS), Microsoft, and Google Cloud and established Common Controls for AI Services—"A collaborative effort to define standardized open-source technology-neutral controls for safe and compliant AI adoption in the financial industry." *This global collaboration reflects growing recognition across the financial ecosystem that proprietary or fragmented approaches are insufficient to address the shared challenges posed by AI adoption in regulated markets. The* **Common Controls for AI Services** *initiative offers a unified framework to drive consistency, transparency, and trust.*[153]

Additional detail is available in Appendix II: Major AI Resources. If banking is of interest to readers, additional research and discussion with experts is advised.

1.11.4.9   Retail—Scan and Go

In April 2024, Sam's Club, the Walmart membership store began implementing a Check Out free customer experience. Their 'Scan and Go' application enables members to scan products as they picked, thus forgoing the traditional checkout lines. The company's intent is to enhance operational efficiency and free associates for better customer engagement.[154]

Time will tell if this is in the retail future, but it is a first step toward dramatically changing the shopping experience since Self-Checking. One critical issue is customer (change management) acceptance as is the model requires significant transformation from the way people have shopped for more than a century.

1.11.4.10   ChatGPT vs Students

According to one college professor, "When students come to school, college, or university, we're not just teaching them how to write, we're teaching them how to think—and that's something no algorithm can replicate." In a research (article published in 2025) from the University of East Anglia (United Kingdom), essays written by 145 students were compared with those generated by ChatGPT. Researchers found that while the AI produced essays were "generally well-structured and grammatically correct, they lacked a key element: the human perspective." AI's staccato speech pattern showed a lack of personal insight and nuance. These findings suggest that academia concerns regarding AI paper submissions are not as visible as some thought. However, this may change as AI software continues to mature.[155]

Subsequently (2025), ChatGPT customized it,

Tone and style can make it more personalized and engaging for your audience. By using smart prompting and easy customization features, you can shape ChatGPT to reflect your unique voice and personality. This can help create a more authentic and relatable interaction with your readers or users. Experimenting with different settings and prompts can help you find the right balance to achieve the desired tone and style. Take advantage of these customization options to make ChatGPT truly your own.[156]

Likely, this process will continue to improve. The professor's tasks will get more difficult at first; however, AI Detection solutions (Section 1.2.1.2 'Deep Learning') will advance as well.[157] This process may mirror the cybersecurity perpetual fight with criminals.

1.11.4.11   Torque Clustering

Physicists define

> Torque is the rotational counterpart of force. Suppose a body rotates about an axis and a force **F** is applied some distance **r** from the axis. The distance from the rotation axis to the point at which the force is applied is called the moment arm. If the force is applied perpendicular to the moment arm, then torque τ is defined as T = Fr.[158]

Taken from physics (the torque balance in gravitational interactions when galaxies merge—properties of mass and distance), "Torque Clustering can efficiently and autonomously analyse vast amounts of data in fields such as biology, chemistry, astronomy, psychology, finance and medicine, revealing new insights such as detecting disease patterns, uncovering fraud, or understanding behaviour."[159]

Torque Clustering outperforms traditional unsupervised learning methods. It has been rigorously tested on over 1,000 diverse datasets averaging a mutual information (AMI) clustering score of 977%. Other current methods generally score in the 80th-percentile range.[160]

1.11.5   FUTURE OF AI RESEARCH

Predictive modeling is in the AI wheelhouse. Prognostication about future developments remains lonely task of human researchers, evangelists, pundits, and others. Water cooler experts aside, the range of possibilities and timing is seemingly endless. In this section, we explore several relevant actions and processes that will drive future AI development.

This list is not all inclusive but simply provides reader with a snapshot of ongoing efforts. It is almost given that the body of knowledge in this area will expand rapidly. These initiatives and actions may set the tone for those that follow.

1.11.5.1   AAI 2025 Presidential Panel

In March 2025, the Association for the Advancement of Artificial Intelligence released its AAI 2025 Presidential Panel on the Future of AI Research. It is provided in this book because the 17 topics included represent a clear identification of the trajectory of near-term AI research. This is important because research is focused on and funded by needs identified from academia as well as practitioners/users and government agencies.[161]

This report is 89 pages in length; therefore, only the **Key Points** are captured in this summary. Each section has a **Context and History** section as well as **Current State and Trends**, **Research Challenges**, and the AAAI **Community Opinion**. Each contains a wealth of information and the positions taken are well documented and cited.

With minor, noted expectations, the following 17 bullets are taken from the referenced (cited) AAI report.

1. **AI Reasoning**—As noted in the Introduction (Section 1.3 'Reasoning'),
   reasoning is core to human intelligence and forms the basis of knowledge.
   Research has led to a range of automated reasoning techniques including
   probabilistic graphical models the pay critical roles in real-world applications. LLM has made significant advancement and now research focus on
   greater correctness and depth is critical, especially for autonomously operating AI agents.

2. **AI Factuality and Trustworthiness**—This is the single largest topic of
   AI research today. Improving "Factually of AI systems based on neural-
   network large language models is at the top of AI research priorities."
   Trustworthiness includes human understanding, robustness, and the incor-
   poration of human values into AI systems. "Approaches to improving the
   factuality and trustworthiness of AI systems include fine-tuning, retrieval-
   augmented generation, verification of machine outputs, and replacing complex models with simple understandable models."

3. **AI Agents**—"Agents and multi-agent systems (MAS) have evolved from
   autonomous problem-solving entities to integrating generative AI and
   LLMs, ultimately leading to cooperative AI frameworks that enhance
   adaptability, scalability, and collaboration."

4. **AI Evaluation**—In AI and Expert Systems: Early Myths, Legends and
   Facts (Section 1.11.2), researches of that era believe software Independent
   Validation and Verification (IV&V) methods for AI systems was similar/the
   same as general software. In this report, the position is taken that this is no
   longer correct. Current benchmarking pays, "Insufficient attention paid to
   other critical factors such as usability, transparency, and adherence to ethical guidelines." New insights and method are needed to assure trustworthiness at wide-scale enterprise deployments.

5. **AI Ethics and Safety**—AI is rapidly becoming mainstream; therefore,
   issues around ethics and safety are becoming paramount. Additionally,
   traditional social and legal frameworks are slow to develop, yet AI-driven
   cybercrime and autonomous weapons require immediate attention. These
   are interdisciplinary and require collaboration and clear AI governance
   (Section 1.7 'Governing AI').

6. **Embodied AI**—"Intelligence emerges through the coupling of a physical
   body with a real environment. Embodied AI insists that coupling is essential to achieving real intelligence in situated agents. Robots are good scientific and engineering platforms for developing Embodied AI."

7. **AI and Cognitive Science**—"Cognitive Science is a multidisciplinary field
   that was inspired by AI's exploration of the hypothesis of computation as
   a scientific language for understanding cognition. Some continued interactions between AI and other areas in cognitive science have yielded valuable
   insights and systems, notably cognitive architecture. Expanding these interactions could yield important advances for both fields."

8. **Hardware and AI**—Computing hardware plays a critical role as its design needs to be optimized for AI high-performance algorithms. Energy and throughput are key challenges when training large-scale models. Hardware of this nature has high energy requirements and requires significant thermal dissipation capability. These limiting factors are challenges for certain applications, including real time.

9. **AI for Social Good**—AI for Social Good (AI4SG) has grown significantly prioritizing ethical considerations. Successful AI4SG initiatives require close collaboration between domain experts, policy makers, and communities as well as the global research community. While significant value can be attained, as with other AI issues, scaling remains a challenge.

10. **AI and Sustainability**—AI computing requires major electricity and water supplies. Innovation is enhancing hardware and software efficiency; however, its value needs to exceed the investment required and operational costs. Moreover, the public must 'perceive' the value (with a lower carbon footprint) if AI is to have long-term sustainability at the scale pundits currently imagine and project.

11. **AI for Scientific Discovery**—AI enables scientific research at an unprecedented pace and new ways of imagining the technological future. With the role of humans changing in this process, ethical issues as well as interdisciplinary collaborative oversight of AI models is required.

12. **Artificial General Intelligence (AGI)**—Human-level capabilities have long been the goal of AI researchers and practitioners—see Turing Test, (Section 1.11 'AI Context from the Perspective of History'). AGI approximates or replaces humans in some equation with potential societal disruptions with associated risks and safety. More research is needed for the double-edged AI sword.

13. **AI Perception vs Reality**—Similar to the Gartner Hype Cycle, the Dunning–Kruger Effect (Section 4.1.1 'D-K Quadrants' of Chapter 4) are cognitive bias tools that depict the tendency of many to 'jump on a technological bandwagon,' and believe new solutions are more mature than they are. Their perception of reality does not match the real-world current situation. Setting realistic expectations can be exceedingly difficult and remains challenging.

14. **Diversity of AI Research Approaches**—By definition (see Glossary), research is the investigation of the new and unknown. Usually, competing hypotheses emerge that must be tested and supported or not supported. There is the danger of Group Think when major segments of the global research community is moving lockstep toward singularly defined goals. We have seen this in other fields, i.e., COVID-19 where so called 'Scientific Consensus' was underwritten by governments and competing theory were held in distain. Cultural Cognition (propensity to perceive risk from one's bias) can play a role as well. Diverse thinking in research leads to new breakthroughs and with emerging technologies such as AI, broad thinking is mandatory.

15. **Research beyond the AI Research Community**—AI is not simply a technical or engineering discipline. AI is societal by nature incorporating cognitive constructs such as ethics, culture, and economics. Seamless collaboration between AI researchers and experts from a very diverse set of fields will assure the best, most effective, and efficient AI solutions with all the benefits therein.

16. **Role of Academia**—Big tech is leading the AI charge. The costs are enormous, and universities cannot compete from financial as well as the depth and breadth of expertise represented in the private sector. Challenges remain as universities seek their long-term roles in this and other rapidly evolving fields.

17. **Geopolitical Aspects and Implications of AI**—In his book, The Rise and Fall of Great Powers: Economic Change and Military Conflict from 1500 to 2000, Paul Kennedy identifies key reasons for changes in relative global power.
    - "Technological and organizational innovation are keys causes in the changes of this relative power.
    - The winners of major wars are generally those that can mobilize the most economic resources.
    - States gain their military power from economic development, but they must be careful not to spend too much of those resources on the military or go to war with too many enemies at once."[162]

AI has taken center stage in much of the geopolitical discussions. However, "Like any category of technology, AI has limits and understanding where those limits lie is crucial to understanding how it might shape global affairs in the near future."[163]

Tensions among competing states will likely continue and universities can have an active role to play through their public policy colleges.

1.11.5.2 AI Convergence

AI is driving a strong convergence among information technologies, psychology (scientific study of the mind and behavior), and scientific/engineering disciplines. While software has always depended on domain expertise to solve high-value problems, AI is somewhat unique in tight linkages, creating a holistic model of human knowledge going forward. Research collaboration between academia and industry are necessary to realize the full value of AI as well as compete with geopolitical knowledge, economic and nation agendas.

Figure 1.6 depicts AI as the convergence of all human knowledge and will be the catalyst for future knowledge growth. It is built upon the RBC framework (Section 2.3 'Relationships, Behaviors, and Conditions' of Chapter 2). The Relationships, Behaviors, and Conditions model is a foundational level of understanding of human interactions. We believe it is very relevant as AI seeks to emulate and perhaps surpass our interactions with each other.

RBC is inherent to AI models.

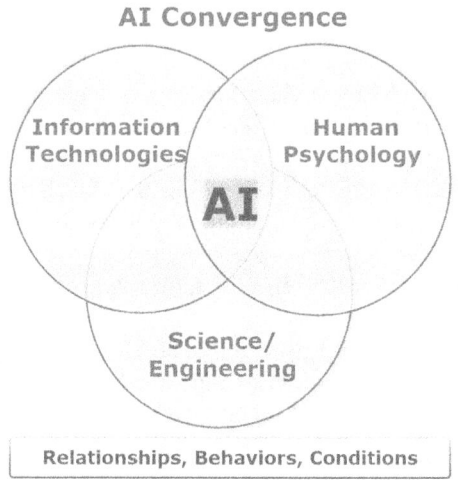

FIGURE 1.6   AI convergence.

There is a demand for significant research and university systems as well as private firms will need to continue and enhance the existing relationships. Likely, this demand will grow and perhaps quickly.

Despite research challenges as articulated in this appendix, AI research is at the dawn of explosive growth. Like the moon shot of the 1960s, this is a great place for those inclined and touches all aspects of humanity and its thirst for knowledge.

1.11.5.3   Systemic Bias

One definition of science, based on the Latin *scientia*, is 'knowledge' or from its present participle of *scire*, 'to know.'[164] This makes the information in this section is disappointing to this degreed physicist. Moreover, we have recognized this concern and have previously addressed it several times.

What is so disturbing about these issues with COVID research appears to be the breadth and depth of poor science. By one count, the publish or perish academic system published "a large volume of misinformation disguised as science during a pandemic obscures true and useful knowledge. At worst, this can lead to bad public health practice and policy."[165]

Beyond lousy scientific method is the intimidation that certain parties including federal health officials perpetrated on those scientists with honest difference of option and/or research findings to the contrary of conventional/accepted wisdom.

In my opinion, this behavior is not science. It is the weaponization of one's scientific thinking. Time to market has decreased for all manner of products and minimum viable product (MVP) is generally accepted, yet fraught with risks as this rush to market accelerates.[166]

Such methodological problems were likely overlooked in the considerably shortened peer-review process for COVID-19 papers. One study estimated the average time from submission to acceptance of 686 papers on COVID-19 to be 13 days, compared with 110 days in 539 pre-pandemic papers from the same journals. In my study, I found that two online journals that published a very high volume of methodologically weak COVID-19 papers had a peer-review process of about three weeks.[167]

The major question for the AI community, what if this behavior becomes part of the development of possible thinking machines? Then ingrained in future AI releases.

1.12   THE UN-MEASURABILITY PROBLEM

According to physicists, Dark Matter and Dark Energy make up the preponderance of our universe; however, we cannot visually see it or measure it directly. Its existence is inferred based on the output of measuring devices that we have some control over. Yet they account for perturbations in our visible universe. Other scientific variables that cannot be measured directly. Examples where inference is made include Hess's law when assessing chemical enthalpy change and Heisenberg's uncertainty principle in physics.[168]

The world is full of variables whose values cannot be directly measured for a number of reasons. These variables impact on all of us, daily. For example, we take relationships for granted but we will show in Section 4.3.1.2 'Variable Taxonomy' of Chapter 4 and throughout the book that we cannot measure relationships directly. We will also discuss the difficulties measuring human behavior using psychometrics, statistical methods useful when measuring behavior and social sciences–perhaps by extension to AI behaviors as well.

Likewise, we have all heard of and may have even experienced the Law of Unintended Consequences, whereby the effects of decisions are unanticipated, unintended, and often not desirable. These may fall into the Unknown Unknowns.

1.12.1   INTELLIGENCE

The scientific study of intelligence is exploring more domain than ever. This includes human, animal, artificial, and even plant intelligence. Intelligence is extended to include not just the cellular but subcellular components.[169]

Often called the father of American psychology William James (1842–1910) is credited with the quotation, "Intelligence is the ability to reach the same goal by different means." He espoused two main schools of thought.

- **Pragmatism**—The concept that the truth of an idea can never be proven, but instead, focus should be on the usefulness of the idea.
- **Functionalism**—The evaluation of mental consciousness (perception, memory, feelings, etc.) as a function of how well it serves the adaption of organisms to their environment (**Conditions**).

"James defined psychology as the conscience of the mental life because he thought that consciousness is what makes mental life possible. He sought to discover the utility of human consciousness and how it is fundamental to survival." He was the first to coin the phrase "stream of consciousness," recognizing that "is nothing jointed; it flows." His belief in free will (indeterminism or unpredictability), also defined as the *degree of possibility*, makes life worth living and holds us all accountable for our actions (**Behaviors**). Finally, his work on Instinct Theory focused on evolution intrinsic behaviors that in addition to learned behaviors are formative of **Relationships**.[170] These are foundational components of the RBC Framework developed in Section 2.3 of Chapter 2.

The word *teleology*, "Comes from two Greek words: telos, meaning 'end, purpose or goal,' and logos, meaning 'explanation or reason.' From this, we get teleology: an explanation of something that refers to its end, purpose or goal."[171]

In 2022, Michael Levin of Tufts University's Allen Discovery Center posited his Technological Approach to Mind Everywhere (TAME)—a framework for understanding and manipulating cognition in unconventional substrates. He argues,

> Synthetic biology and bioengineering provide the opportunity to create novel embodied cognitive systems (otherwise known as minds) in a very wide variety of chimeric architectures combining evolved and designed material and software. These advances are disrupting familiar concepts in the philosophy of mind and require new ways of thinking about and comparing truly diverse intelligences, whose composition and origin are not like any of the available natural model species.

He goes on to state that AI can be considered one of many diverse cognitive agents.[172]

One of his summary statement is indicative in this writer's belief that we are rapidly if not there already, able to investigate AI similarly as we study human and other animal intelligence.

> Intelligence is the degree of competency of navigating any space (not just the familiar 3D space of motility), including morphospace, transcriptional space, physiological space, etc., toward desirable regions, while avoiding being trapped in local minima. Estimates of intelligence of any system are observer-dependent and say as much about the observer and their limitations as they do about the system itself.[173]

Additionally, unconventional intelligences are contrary to our expectation about intelligences. They are often not brain-based and may not even be alive in the traditional sense. One example is the *motor behavior* associated with animal muscles. *Degrees of freedom* suggests that there are a large number of combinational possibilities for accomplishing a task requiring physical activity. The largeness of these possibilities may overload the brain.

This suggests that components (muscles, nerves, etc.) seek the stated goals and navigates potential obstacles and perturbations, freeing the brain for higher level functionality. An example, babies solving the problem of walking in diapers. Learning associated with this unnatural requirement to walk is intelligence. This morphogenesis, "Is the classic example of unconventional intelligence, and the associated models take inspiration from Alan Turing's analysis of pattern formation in chemical reactions."

Muscle motor behaviors navigating toward a goal (optimal) state without neural instruction are unconventional intelligence.[174] Sometimes, colloquially referred to as *Muscle Memory*, an automatic, unconscious movement such as a skill in which individuals are trained.[175]

Distributed computing models come to mind. This is a method whereby multiple computers work together to solve problems, and the network appears to be a single large-scale resource addressing significant complexity. This type of network can deal with big data and sets of differential equations as well as high resolution graphics.[176] Sound familiar?

This AI Convergence of human and non-human intelligences is further discussed in several Sections throughout the book. AI is expanding science into new realms, as reviewed in Section 1.11.4.5 'Urania—AI Sensor Technology Breakthrough.' These high-end AI algorithms are just a few of those that will continue to exponentially expand human horizons in the coming decades.

One final point which will be developed in more detail comes from a Substack writer. He is one of many who are expressing concern about AI, its ethics, job threat, etc. He argues that at the end of the day, AI is still a machine. He argues that in the hands of unethical or even evil individuals, great damage can be done. His fears are well founded, and life-changing technologies can do great damage in the wrong hands, i.e., nuclear energy, cyber, aviation, explosive chemicals, etc. Societies have always been putting checks and balances in place, and this technology will be no different.

1.12.2   THE LATENT CONSTRUCT

Latent variables are an essential concept in statistics, machine learning, and various scientific disciplines, particularly in areas involving complex data analysis and modelling. Unlike observable variables, which can be directly measured or observed, latent variables represent underlying factors or constructs that are not directly observable but are inferred from other measurable variables.[177]

Validity involves two different aspects: content validity and constructs validity.

1. The content validity shows the degree of concordance among a panel of specialists and evaluates if the items are representative of the domain that the scale will evaluate.
2. The constructs validity implies that the scale measure what was purported to be measured. The constructs validity can be **convergence** (items that make up the construct are correlated); **discriminant** (some items in the construct do not correlate with others in the construct); or **criterion** (the operationalization of a construct agrees with previously established criteria).[178]

These types of variables can be addressed using psychometric measurement scales are affected by the variables that comprise them.[179] This can insert unknown bias into the assessment and statistical model require certain assumptions be made. We don't want to get too far in the weeds, but it is important that management understand this is the variable universe AI/ML live in. Appropriate mathematicians and SMEs (possibly including behavioral experts) must be engaged in the development process.

It is useful to briefly describe some of the scientific and mathematical foundational knowledge base that preceded AI. Indeed, AI inference is built on these foundations.

- **Behavioral Science**—"A branch of science (such as psychology, sociology, or anthropology) that deals primarily with human action and often seeks to generalize about human behavior in society."[180]
- **Game Theory**—"The study of the ways in which interacting choices of economic agents produce outcomes with respect to the preferences (or utilities) of those agents, where the outcomes in question might have been intended by none of the agents."[181]
- **System Dynamics**—"A computer-aided approach for strategy and policy design. The main goal is to help people make better decisions when confronted with complex, dynamic systems. The approach provides methods and tools to model and analyzes dynamic systems. Model results can be used to communicate essential findings to help everyone understand the system's behavior. It uses simulation modeling based on feedback systems theory that complements systems thinking approaches. It applies to dynamic problems arising in complex social, managerial, economic, or ecological systems. It can be applied to social, managerial, economic, ecological, and physiological systems."[182]

Figure 1.7 depicts the relationship between the foundational forces. Note that the nexus is a Latent Construct.

Each is discussed at a high level and integrated into the construct of Structural Dynamics. The AI construct, Torque Clustering is built upon this broad body of knowledge. The commonality of these foundational elements is their assessment of variables and their relationships that are not intuitive or even visible, even to the trained scientist/observer.

This fuzziness is a fundamental cause of uncertainty and thus inherent error(s) of traditional mathematical solutions based on potentially biased observations, assessments,

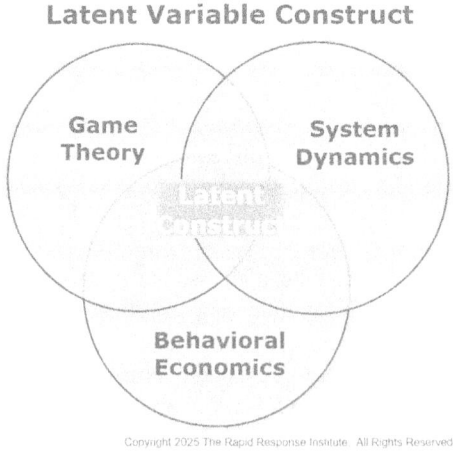

FIGURE 1.7   Latent variable construct.

and conclusions. Moreover, those involved in this assessment or decision-making process may fall prey to Observer Bias, known by psychologist as Detection or Ascertainment Bias where preconceptions and expectations cloud the analysis.[183]

1.12.2.1 Behavioral Science

We, physicists, believe that physics is the foundational science of everything. More objective minds state, "Physics can, at base, be defined as the science of matter, motion, and energy. Its laws are typically expressed with economy and precision in the language of mathematics."[184]

Therefore, it is no surprise that the mathematics of *torque* (Section 1.11.4.11 'Torque Clustering') describes certain AI analytic process behaviors. Moreover, it is not a big step to understand *human behavior* mathematically using the scientific method.

The study of human behavior using systematic experimentation and observation is referred to as Behavioral Science. The field of behavioral economics explains and predicts behaviors *not anticipated by standard economic theories.*[185]

Behavioral science began in ancient Greece when philosophers such as Aristotle mused about human behavior and its causes. Behavioral science as we know today did not start evolving until late in the 19th century with the establishment of the first psychology laboratory by Wilhelm Wundt in 1879. Key disciplines include the following:

- **Psychology**—Scientific study of the mind and behavior
- **Sociology**—Study of societies and its forces that shape behavior
- **Anthropology**—Study of the cultural aspects of human behavior
- **Economics**—Study of the combination of economics and psychology in the human (economic) decision-making process

A few of the most influential theories that have emerged include the following:

- **Maslow's Hierarchy of Needs**—Human behavior as a function of need from survival to self-actualization (Section 2.5 'Maslow's Hierarchy of Needs' of Chapter 2)
- **Pavlov's Classical Conditioning**—Learned behavior resulting from conditioning
- **Social Learning Theory**—individual can acquire new behaviors by imitating actions of others by observation
- **The Stanford Prison Experiment**—The influence of human behavior by situational factors

These and other theories have resulted in a wide and growing range of applications, in the fields marketing, public policy, education, mental health, and others.[186] Specific Marketing (Section 4.3.2.1 'Market Assessment' of Chapter 4) and Education (Section 4.2.9 'Role of Humans in the AI Era' of Chapter 4) AI models will be addressed herein.

1.12.2.2 Game Theory

In 1994, the Nobel Prize in Economic Sciences was *awarded jointly to Professor John C. Harsanyi, University of California, Berkeley, CA, USA; Dr. John F. Nash, Princeton University, Princeton, NJ, USA; and Professor Dr. Reinhard Selten, Rheinische Friedrich-Wilhelms-Universität, Bonn, Germany for their pioneering analysis of **equilibria in the theory of non-cooperative games**.*[187] In non-cooperative games, players pursue their own interests which are effectively real-world scenarios.[188] While players can work with each other to address a human interaction problem, their ultimate goal is their personal success in the negotiation situation. We see that in B2B sales, where the goal is to 'make a deal,' yet each side seeks to maximize its own value. Political wrangling is another example of this process.

One of the more interesting aspects of non-cooperative games is the concept of *equilibrium*. It turns out that this game phenomenon has *many stable outcomes*, or multiple equilibria which is an issue when making economic or behavioral predictions.[189]

> The Theory of Non-Cooperative Games studies and models conflict situations among economic agents; that is, it studies situations where the profits (gains, utility or payoffs) of each economic agent depend not only on his/her own acts but also on the acts of the other agents.

Non-Cooperative Game Theory is very useful for modelling and understanding multi-personal economic problems characterized by strategic interdependencies.[190]

We will discuss game theory more fully in Section 6.6 'Case Five—AI-Driven Cross-Cultural Serious Gaming' of Chapter 6. In that section, we will review the process currently underway to re-formulate an existing online training game to AI in lieu of human players.

This process is built from the IBM chess playing AI solution from the 1990s, Deep Blue. This supercomputer enable AI solution could explore up to 200 million possible chess positions per second. The game is well known and played and the rules understood by millions. According a 2017 Scientific American interview, "It's known as a game that requires strategy, foresight, logic—all sorts of qualities that make up human intelligence. So, it makes sense to use chess as a measuring stick for the development of artificial intelligence." To put things into perspective, IBM computer have been playing chess since the late 1940s.[191]

A 2025 research paper analyzed almost 32,000 prose rationales and suggested that LLMs cooperative gaming resulted in likely opponent strategy in a timely manner. **This work connects classic game theory with machine psychology, offering a rich and granular view of algorithmic decision-making under uncertainty.**[192]

1.12.2.3 System Dynamics

System Dynamics Modeling is a methodology for analyzing and understand complex models. It was created by Jay W. Forrester at MIT in the mid-1950. The model simulates processes through iterative flow using stocks and flows with feedback and causal relationships among model variables.

In our 2025 book, *Navigating the Data Minefields: Management's Guide to Better Decision-Making*, we presented in depth the use of System Dynamics to address CAPEX Risk. Interested readers should review that use case put together for a large oil and gas firm that was concerned about the challenges inherent to 'difficult' and expensive oil well drilling programs.[193]

The full Drilling Risk Assessment model was presented in detail. This approach to risk management was initiated to support risk assessment in difficult oil well drilling environments. We have used similar versions for other risk assessment in critical infrastructure sectors, and it is a very useful tool, although expertise in its software is required.

1.12.3 STRUCTURAL DYNAMICS

In the 1990s, the author first integrated two methods, Game Theory and Structural Equation Modeling extending the Relationships, Behavior, and Conditions (RBC) Framework systemically (structure and process).[194] This methodology was codified in our 2015 book, *Structural Dynamics: Foundation of Next Generation Management Science*.[195]

The following is taken from a 2016 blog where we addressed issues around the election surprise of that year.[196] Flash forward to 2024, and the same thing happened. In Section 2.5, we develop an AI model designed to help us better understand these hidden (latent) processes.

First the United Kingdom BREXIT vote and more recently the United States 2016 election cycle.

Why does this continue to happen? In an era of sophisticated polls, unlimited campaign spending, and Big Data analysis, not just a few but most missed these social earthquakes.

Some are now suggesting that the President Elect had tapped into something that the defeated party did not see—and many of his own for that matter! If this is so, how did this individual recognize this tsunami when the political pros on both sides could not?

We started researching this phenomenon in the early 1990s. Following primary efforts, "integrating structural and process components into a dynamic system model" as part of systems analysis of human interaction, we coined the term Structural Dynamics.

Other early publications in the form of articles and speeches were formalized into our "Beta" White paper which was released in 2012. In 2015, our first Monograph in Changing the Dialogue: A Series on the New Business Dynamics formally documented Structural Dynamics: Foundation of Next Generation Management Science.

We have defined Structural Dynamics as, "The morphology or patterns of motion toward process equilibrium of interpersonal systems." This is founded in the hypothesis that structure and process are intertwined and often the latent or unseen variables only manifest themselves at a (now visible) tipping point.

Structural Dynamics

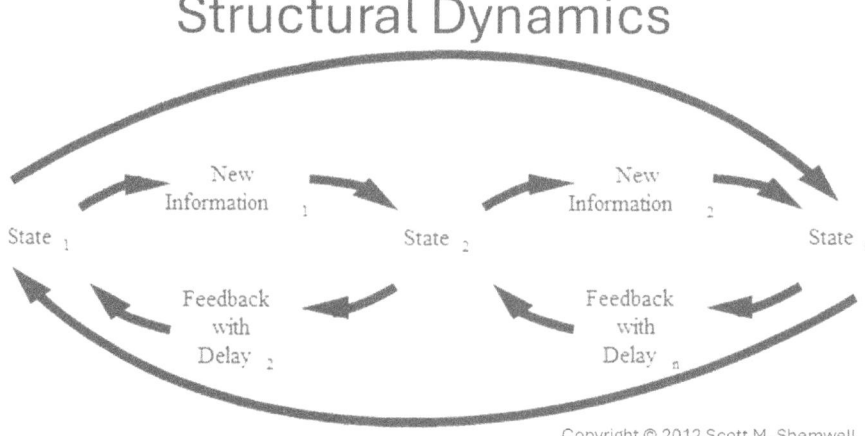

FIGURE 1.8   Structural dynamics.

The following mathematics describes the process depicted in Figure 1.8. The current state of a system ($State_n$) is a function of all new information (NI_n) which is the sum of current information (CI_n) plus feedback (F_n).

$$State^n = \sum (NI_{1...n})$$

where

$$NI_{1...n} = (CI_{1...n} + F_{1...n})$$

Equation 1.1—Structural Dynamics State Change

Moreover, system delay can be either by design or through system imperfections. Delays can be caused by the system or individual's initial lack of understanding of the impact of the new information, flaws in the feedback mechanisms, inaccurate or incomplete analysis of the new data, or confusion caused by the interaction with other systems.

New information process iteration based on **Behaviors** resulting in a New State or **Condition**. The end result of this model is a new **Relationship(s)** between parties. New States can also represent changes in Maslow's Hierarchy (Section 2.5 of Chapter 2).

The resulting, Cross-Cultural Serious Game and its applicability to AI will subsequently be developed in Section 6.6 of Chapter 6. We continue to find broad use of the methodology and believe AI can help take it to the next level.

1.12.3.1   Structural Equation Modeling

Structural Equation Modeling (SEM) is a challenging and mathematics intensive tool that can be used to assess latent variables. It is not the intention of this book to address such a complicated but to make executives away how AI uses this type of statistical tool. As with other examples herein, it is imperative that someone in the firm have a working knowledge of SEM and the nature of the problem it is used to address.

> Many important attributes in the social, behavioral, and health sciences cannot be observed directly. Examples of such attributes include happiness, depression, anxiety, cognitive and social competence, etc. They are typically measured by multiple indicators that are often subject to measurement errors. Structural equation modeling (SEM) has become a major tool for examining and understanding relationships among latent attributes.[197]

SEM will be referred to again in Section 6.4.2.2 'Planned Enhancements' of Chapter 6.

1.13 MISCELLANEOUS AI SELECTION CRITERIA

In Section 1.2.1 'Major Application Models,' we described the major AI models in use. Readers should review that listing as part of the technology assessment and selection process. In this section, we will look at some of the other issues management faces when making AI implementation decisions.

1.13.1 MODEL CONTEXT PROTOCOL

> MCP is an open protocol that standardizes how applications provide context to LLMs. Think of MCP like a USB-C port for AI applications. Just as USB-C provides a standardized way to connect your devices to various peripherals and accessories, MCP provides a standardized way to connect AI models to different data sources and tools.[198]

This is another issue that does not get much press that your AI development team must know, understand, and use as appropriate.

1.13.2 DECENTRALIZED AI

Historically, IT systems start as centralized large scale computing solutions. The old IBM Mainframes ruled for years before being replaced minicomputers and ultimately the personal computer and smart devices of today. Expect AI to follow a similar trajectory.

At a high level, decentralized AI looks similar. Data stays on the local server and training is conducted locally (federated learning), thus providing additional control and privacy of sensitive information. Moreover, blockchain acts as the digital ledger.

> Instead of a single entity deciding AI updates, multiple devices (or nodes) must reach an agreement before any changes take effect. This is done through consensus mechanisms like voting or staking, ensuring that AI evolves fairly and transparently. Since no single company or organization has full control, decentralized AI is more democratic, resistant to censorship, and aligned with user interests.[199]

Much of what is discussed in Chapter 6 'Capstone—Detailed AI Models under Development' would meet the test of decentralized AI. Expect decentralized AI to play a larger role in everyday AI life.

1.13.3   IP OWNERSHIP

This section is not legal advice, and the author is not an attorney. This information is educational purposes only and legal counsel should be consulted, as necessary.

We have discussed ethics, legal, and risk issues herein. Always consider whether the data being accessed by AI is a copyright product of another entity or individual. This is true for images as well, whether art or human likeness. Always run any AI product or solution by legal before release for assurance that there is no infringement on the rights and ownership of others. This release should be written into AI governance as well.

NOTES

1. "The Only Rules Are the Ones Dictated by the Laws of Physics." goodreads. https://www.goodreads.com/work/quotes/144038357-elon-musk. Accessed August 12, 2025.
2. "Dotcom Bubble Definition." Investopedia. https://www.investopedia.com/terms/d/dotcom-bubble.asp. Accessed March 9, 2025.
3. "Tulips." Governing Energy. https://therrinstitute.com/wp-content/uploads/2018/10/Tulips-March-18-2013.pdf. Accessed March 10, 2025.
4. "Great Depression." Britannica. https://www.britannica.com/event/Great-Depression. Accessed August 6, 2025.
5. "The 2008 Financial Crisis Explained." Investopedia. https://www.investopedia.com/articles/economics/09/financial-crisis-review.asp. Accessed August 7, 2025.
6. "Management Theory—Evolution Not Revolution." Scott M. Shemwell. https://therrinstitute.com/wp-content/uploads/2019/10/1993-Management-Theory-Evolution-Not-Revolution.pdf. Accessed July 23, 2025.
7. "The History of Artificial Intelligence: Complete AI Timeline." TechTarget. https://www.techtarget.com/searchEnterpriseAI/tip/The-history-of-artificial-intelligence-Complete-AI-timeline. Accessed July 23, 2025.
8. "Types of AI Models Explained: A Deep Dive into Architectures and Applications." Geniatech. https://www.geniatech.com/ai-models-types-explained/. Accessed July 24, 2025.
9. "Types of AI Models Explained: A Deep Dive into Architectures and Applications." Geniatech. https://www.geniatech.com/ai-models-types-explained/. Accessed July 24, 2025.
10. "A List of Large Language Models." IBM. https://www.ibm.com/think/topics/large-language-models-list. Accessed October 7, 2025.
11. "Types of AI Models Explained: A Deep Dive into Architectures and Applications." Geniatech. https://www.geniatech.com/ai-models-types-explained/. Accessed July 24, 2025.
12. "Understanding the Different Types of Artificial Intelligence." IBM. https://www.ibm.com/think/topics/artificial-intelligence-types. Accessed July 24, 2025.
13. "The Limits of Reason." Psychology Today. https://www.psychologytoday.com/us/blog/hide-and-seek/201902/the-limits-reason?msockid=07617bc0a43661233e436d82a5ce608d. Accessed August 8, 2025.
14. "What Is Reasoning in AI?" IBM. https://www.ibm.com/think/topics/ai-reasoning, Accessed August 9, 2025.

15. "Do LLMs Understand Morality?" Bhavishya Pandit. https://media.licdn.com/dms/document/media/v2/D4D1FAQHmQYSQ4h564g/feedshare-document-pdf-analyzed/B4DZSMVt3BHUAY-/0/1737521303062?e=1755734400&v=beta&t=A7AM4OGU_uRvXRRR_rLYoolgBVNMVTqhEsdrCZQXABk. Accessed July 2, 2025.
16. "Affective Computing." ScienceDirect. https://www.sciencedirect.com/topics/computer-science/affective-computing. Accessed August 30, 2025.
17. "What Is Affective Computing and How Could Emotional Machines Change Our Lives?" Bernard Marr & Co. https://bernardmarr.com/what-is-affective-computing-and-how-could-emotional-machines-change-our-lives/. Accessed August 30, 2025.
18. "Why AI Will Never Truly Understand Your Feelings — And Why That Matters." Forbes. https://www.forbes.com/sites/bernardmarr/2025/04/23/why-ai-will-never-truly-understand-your-feelings-and-why-that-matters/. Accessed August 30, 2025.
19. "Microsoft AI CEO Suleyman Is Worried about 'AI Psychosis' and AI That Seems 'Conscious'." Fortune. https://fortune.com/2025/08/22/microsoft-ai-ceo-suleyman-is-worried-about-ai-psychosis-and-seemingly-conscious-ai/. Accessed August 30, 2025.
20. "Breaking Down the Lawsuit against OpenAI over Teen's Suicide." TechPolicy.PRESS. https://www.techpolicy.press/breaking-down-the-lawsuit-against-openai-over-teens-suicide/. Accessed August 30, 2025.
21. "Wechsler Adult Intelligence Scale–Third Edition." APA PsycNet. https://psycnet.apa.org/doiLanding?doi=10.1037%2Ft49755-000. Accessed July 6, 2025.
22. "Modeling for Learning Organizations." Systems Thinker. https://thesystemsthinker.com/modeling-for-learning-organizations/. Accessed July 8, 2025.
23. "Researchers Claim Their AI Model Simulates the Human Mind. Others Are Skeptical." Science. https://www.science.org/content/article/researchers-claim-their-ai-model-simulates-human-mind-others-are-skeptical?_bhlid=b012eb735596da5010fa02ca03398bed683ecca7. Accessed July 8, 2025.
24. "Are We Truly Rational? AI Challenges a Long-Held Scientific Belief." SciTechDaily. https://scitechdaily.com/are-we-truly-rational-ai-challenges-a-long-held-scientific-belief/. Accessed July 12, 2025.
25. "What Is the Law of Attraction & How Does It Work?" The Law of Attraction. https://thelawofattraction.com/what-is-the-law-of-attraction/. Accessed May 1, 2025.
26. "The 17-Second Rule: The Neuroscience of Focused Intention, Law of Attraction, and Gratitude." The Critical Thought Lab. https://thecriticalthoughtlab.com/the-17-second-rule-the-neuroscience-of-focused-intention-law-of-attraction-and-gratitude/. Accessed May 1, 2025.
27. "What Is Unconscious Bias (And How You Can Defeat It)." Psychology Today. https://www.psychologytoday.com/us/blog/intentional-insights/202007/what-is-unconscious-bias-and-how-you-can-defeat-it. Accessed May 1, 2025.
28. "How Do You Know the App Is Giving You the Right Answer?" The Rapid Response Institute. https://therrinstitute.com/how-do-your-know-the-app-is-correct/. Accessed May 1, 2025.
29. "How to Identify Bias: 14 Types of Bias." MasterClass. https://www.masterclass.com/articles/how-to-identify-bias?campaignid=486026846&adgroupid=1241349711708351&adid=77584490928212&source=Bing&medium=Search&campaign=%5BMC%5D%20%7C%20Search%20%7C%20NonBrand%20%7C%20All%20Website_DSA%20Consolidated%20%7C%20ALL%20%7C%20EN%20%7C%20tCPA%20%7C%20EG&gclid=75665e2fe22e132bd870361995147202&gclsrc=3p.ds&msclkid=75665e2fe22e132bd870361995147202. Accessed May 29, 2025.
30. "How to Identify Cognitive Bias: 12 Examples of Cognitive Bias." MasterClass. https://www.masterclass.com/articles/how-to-identify-cognitive-bias?_gl=1*87p6ii*_gcl_dc*R0NMLjE3NDg0MDA5MjkuNzU2NjVlMmZlMjJlMTMyYmQ4NzAzNjE5OTUxNDcyMDI.*_gcl_au*MzcyODk3Nzk4LjE3NDgzOTk5OTQ. Accessed May 30, 2025.
31. "13 Types of Common Cognitive Biases That Might Be Impairing Your Judgment." Verywellmind. https://www.verywellmind.com/cognitive-biases-distort-thinking-2794763. Accessed May 29, 2025.

32. "Five Types of Cognitive Bias in Data Science (And How to Avoid Them)." builtin. https://builtin.com/data-science/cognitive-biases-data-science. Accessed May 29, 2025.
33. "Data Bias: The Latent or Unobserved." The Rapid Response Institute. https://therrinstitute.com/data-bias-the-latent-or-unobserved/. Access May 29, 2025.
34. "Quote Origin: There Are Three Kinds of Lies: Lies, Damned Lies, and Statistics." Quote Investigator. https://quoteinvestigator.com/2022/06/22/lies-statistics/. Accessed May 28, 2025.
35. "5 Types of Statistical Bias to Avoid in Your Analyses." Harvard Business School Online. https://online.hbs.edu/blog/post/types-of-statistical-bias. Accessed May 28, 2025.
36. "Statistical Bias: 6 Types of Bias in Statistics." builtin. https://builtin.com/data-science/types-of-bias-in-statistics. Accessed May 29, 2025.
37. "Why Coca-Cola's 'New Coke' Flopped." History. https://www.history.com/articles/why-coca-cola-new-coke-flopped. Accessed May 29, 2025.
38. Shemwell, Scott M. (1996). Cross Cultural Negotiations between Japanese and American Businessmen: A Systems Analysis, (Exploratory Study). Unpublished doctoral dissertation, Nova Southeastern University, Ft. Lauderdale.
39. "Navigating the Data Minefields: Management's Guide to Better Decision-Making." CRC Press. https://www.routledge.com/Navigating-the-Data-Minefields-Managements-Guide-to-Better-Decision-Making/Shemwell/p/book/9781032677934#top. Accessed May 28, 2025.
40. "High Reliability Management in Process Industries: Sustained by Human Factors." The Rapid Response Institute. https://therrinstitute.com/wp-content/uploads/2018/10/HRM-in-Process-Industries-Sustained-by-Human-Factors.pdf. Accessed May 28, 2025.
41. "6 Ways to Combat Bias in Machine Learning." builtin. https://builtin.com/machine-learning/bias-machine-learning. Accessed May 29, 2025.
42. "Welcome to Our Cross-Cultural Serious Game Portal." The Rapid Response Institute. https://therrinstitute.com/lets-play-a-game/. Accessed May 29, 2025.
43. "Computers Know What to Do with 1s and 0s: How So?" techwithtech. https://techwithtech.com/computers-know-what-to-do-with-1s-0s/. Accessed June 9, 2025.
44. "AI's Mathematical Mirage: Apple Study Challenges Notion of AI Reasoning." IBM. https://www.ibm.com/think/news/apple-llm-reasoning. Accessed June 9, 2025.
45. "GSM-Symbolic: Understanding the Limitations of Mathematical Reasoning in Large Language Models." Apple. https://arxiv.org/pdf/2410.05229. Accessed June 9, 2025.
46. "What Is Synthetic Data?" IBM. https://research.ibm.com/blog/what-is-synthetic-data. Accessed June 9, 2025.
47. "Data Contamination." Encyclopedia.com. https://www.encyclopedia.com/computing/dictionaries-thesauruses-pictures-and-press-releases/data-contamination. Accessed June 9, 2025.
48. "Understanding the Different Types of Artificial Intelligence." IBM. https://www.ibm.com/think/topics/artificial-intelligence-types. Accessed June 9, 2025.
49. "Understanding the Different Types of Artificial Intelligence." IBM. https://www.ibm.com/think/topics/artificial-intelligence-types. Accessed June 9, 2025.
50. "The Illusion of Thinking: Understanding the Strengths and Limitations of Reasoning Models via the Lens of Problem Complexity." Apple. https://ml-site.cdn-apple.com/papers/the-illusion-of-thinking.pdf. Accessed June 10, 2025.
51. "Intuition vs. Pattern Recognition." This vs. That. https://thisvsthat.io/intuition-vs-pattern-recognition. Accessed June 20, 2025.
52. "The Explanation of Pattern Recognition Model with Cognitive Psychology." ISPRS. https://www.isprs.org/proceedings/XXXIV/part2/paper/089_138.pdf. Accessed June 19, 2025.
53. "Can AI Understand? The Chinese Room Argument Says No, But Is It Right?" Honest AI. https://honestaiengine.com/can-ai-understand-the-chinese-room-argument-says-no-but-is-it-right?utm_source=ainews.honestaiengine.com&utm_medium=newsletter&utm_campaign=ai-s-emotional-puzzle-what-machines-can-t-comprehend&_bhlid=22f673d046e5acafdb823025fecd17da796e773f. Accessed August 7, 2025.

54. "Integrity Is the Essence of Everything Successful." BrainyQuote. https://www.brainyquote.com/quotes/r_buckminster_fuller_153437. Accessed April 17, 2025.

55. "Asset/Equipment Integrity Governance: Operations–Enterprise Alignment." The Rapid Response Institute. https://therrinstitute.com/wp-content/uploads/2017/10/asset_integrity_governance_-ver_1.1.pdf. Accessed October 3, 2025.

56. "White House Unveils America's AI Action Plan." The White House. https://www.whitehouse.gov/articles/2025/07/white-house-unveils-americas-ai-action-plan/. Accessed July 30, 2025.

57. "Management Theory – Evolution Not Revolution." Scott M. Shemwell. https://therrinstitute.com/wp-content/uploads/2019/10/1993-Management-Theory-Evolution-Not-Revolution.pdf. Accessed April 23, 2025.

58. "Solving the AI ROI Puzzle." IBM. https://www.ibm.com/thought-leadership/institute-business-value/en-us/report/chief-ai-officer?utm_source=www.humanintheloop.online&utm_medium=newsletter&utm_campaign=7-edition-mcp-a-simple-guide-to-10x-your-claude-setup&_bhlid=b6d20cb0f5fed8568dfd709f2dbe28bb38888f8a. Accessed July 24, 2025.

59. "Implementing a Culture of Safety: A Roadmap for Performance Based Compliance." Xlibris. https://www.xlibris.com/en/bookstore/bookdetails/552562-implementing-a-culture-of-safety. Accessed July 25, 2025.

60. "AI-Generated Disinformation in Europe and Africa: Use Cases, Solutions and Transnational Learning." Konrad-Adenauer-Stiftung. https://media.licdn.com/dms/document/media/v2/D4E1FAQHWMMncMp3o3g/feedshare-document-pdf-analyzed/B4EZZ.IZBbHcAc-/0/1745872892550?e=1747267200&v=beta&t=9Hc0rovWr4D4XKpn_zIINCHkSPsT5KlTjFruuS-tV-k. Accessed May 1, 2025.

61. "'Periodic Table of Machine Learning' Could Fuel AI Discovery." MIT News. https://news.mit.edu/2025/machine-learning-periodic-table-could-fuel-ai-discovery-0423. Accessed April 23, 2025.

62. "High Reliability Management in Process Industries: Sustained by Human Factors." Scott Shemwell and Denise Brookes. https://therrinstitute.com/wp-content/uploads/2018/10/HRM-in-Process-Industries-Sustained-by-Human-Factors.pdf. Accessed July 26, 2025.

63. "Insights from Complexity Theory: Understanding Organizations Better." tejas.iimb. https://tejas.iimb.ac.in/articles/12.php. Accessed July 26, 2025.

64. "The Economics of an NFL Franchise: Revenue Streams and Financial Challenges." Frontproof Media. https://www.frontproofmedia.com/football/the-economics-of-an-nfl-franchise-revenue-streams-and-financial-challenges. Accessed April 9, 2025.

65. "4 Leadership Lessons from the NFL's Chief Data Officer." KelloggInsight. https://insight.kellogg.northwestern.edu/article/4-leadership-lessons-from-the-nfls-chief-data-officer. Accessed June 22, 2025.

66. "Urban Artificial Intelligence: From Real-World Observations to a Paradigm-Shifting Concept." Medium. https://medium.com/urban-ai/urban-artificial-intelligence-from-real-world-observations-to-a-paradigm-shifting-concept-27f4c5aa20f6. Accessed June 22, 2025.

67. "What Is a Smart City?" IBM. https://www.ibm.com/think/topics/smart-city. Accessed June 22, 2025.

68. "How a Top Company Used Generative AI to Reinvent Culture and Innovation." Forbes. https://www.forbes.com/sites/dianehamilton/2025/06/07/how-a-top-company-used-generative-ai-to-reinvent-culture-and-innovation/. Accessed June 11, 2025.

69. "How a Top Company Used Generative AI to Reinvent Culture and Innovation." Forbes. https://www.forbes.com/sites/dianehamilton/2025/06/07/how-a-top-company-used-generative-ai-to-reinvent-culture-and-innovation/. Accessed June 11, 2025.

70. "Coca-Cola: The Future Is 'AI Meets Human Ingenuity'." MarketingWeek. https://www.marketingweek.com/coca-cola-artificial-intelligence/. Accessed June 11, 2025.

71. "Predicting the Unpredictable: AI's New Role in Aviation Safety." Honest AI. https://honestaiengine.com/predicting-the-unpredictable-ais-new-role-in-aviation-safety?utm_source=ainews.honestaiengine.com&utm_medium=newsletter&utm_campaign=ai-video-mocks-trump-s-factory-revival-dreams&_bhlid=73f715c58150e69057ab1394275389e2ec279c3e. Accessed April 9, 2025.

72. "Monthly Aviation Dashboard." NTSB. https://www.ntsb.gov/safety/data/Pages/monthly-dashboard.aspx. Accessed April 9, 2025.

73. "Predicting the Unpredictable: AI's New Role in Aviation Safety." Honest AI. https://honestaiengine.com/predicting-the-unpredictable-ais-new-role-in-aviation-safety?utm_source=ainews.honestaiengine.com&utm_medium=newsletter&utm_campaign=ai-video-mocks-trump-s-factory-revival-dreams&_bhlid=73f715c58150e69057ab1394275389e2ec279c3e. Accessed April 9, 2025.

74. "AI in Aviation and Airlines: Use Cases for 2024." MINDTITAN. https://mindtitan.com/resources/industry-use-cases/ai-in-aviation-and-travel/. Accessed April 9, 2025.

75. "United Airlines CEO: 'We're Probably Doing More AI than Anyone'." CIODIVE. https://www.ciodive.com/news/united-airlines-ceo-AI-use-cases/749563/?utm_source=Sailthru&utm_medium=email&utm_campaign=Newsletter%20Weekly%20Roundup:%20CIO%20Dive:%20Daily%20Dive%2006-07-2025&utm_term=CIO%20Dive%20Weekender. Accessed June 11, 2025.

76. "Artificial Intelligence and Machine Learning in Software as a Medical Device." FDA. https://www.fda.gov/medical-devices/software-medical-device-samd/artificial-intelligence-and-machine-learning-software-medical-device. Accessed July 1, 2025.

77. "Introducing HealthBench." OpenAI. https://openai.com/index/healthbench/. Accessed August 8, 2025.

78. "Kierkegaard: Life Can Only Be Understood Backwards, but It Must Be Lived Forwards." Philosophy Break. https://philosophybreak.com/articles/kierkegaard-life-can-only-be-understood-backwards-but-must-be-lived-forwards/. Accessed May 2, 2025.

79. "Operational Excellence Platform. The Rapid Response Institute." https://therrinstitute.com/operational-excellence-platform/. Accessed April 7, 2025.

80. "RRI Software Suite." The Rapid Response Institute. https://therrinstitute.com/software-suite/. Accessed April 7, 2025.

81. "OE Solution Set." The Rapid Response Institute. https://therrinstitute.com/oe-solution-set-2/. Accessed April 7, 2025.

82. "DOGE'd." The Rapid Response Institute. https://therrinstitute.com/doged/. Accessed April 6, 2025.

83. "We Have to Pass the Bill so You Can Find Out What's in It." Texas Public Policy Foundation. https://www.texaspolicy.com/we-have-to-pass-the-bill-so-you-can-find-out-whats-in-it/. Accessed April 6, 2025.

84. "Doge: Department of Government Efficiency." https://doge.gov/savings. Accessed April 7, 2025.

85. "The Impact of AI on Consulting Hiring." StrategyCase.com. https://strategycase.com/the-impact-of-ai-on-consulting-hiring/. Accessed April 7, 2025.

86. "Scientific Management and the Knowledge Worker of the 1990's." Proceedings of the 11th Annual Association of Management Conference. https://therrinstitute.com/wp-content/uploads/2019/10/1993-Scientific-Management-and-the-Knowledge-Worker-of-the-1990s.pdf. Accessed April 7, 2025.

87. Andreas Horm. LinkedIn. https://www.linkedin.com/in/andreashorn1/. Accessed August 23, 2025.

88. "Generative AI Customer Stories." Amazon. https://aws.amazon.com/ai/generative-ai/customers/. Accessed August 24, 2025.

89. "Harnessing the Value of Generative AI: 2nd Edition: Top Use Cases across Sectors." Capgemini. https://www.capgemini.com/wp-content/uploads/2024/05/Final-Web-Version-Report-Gen-AI-in-Organization-Refresh.pdf. Accessed August 24, 2025.

90. "The AI Dossier: A Collection of Our Latest High-Impact AI Use Cases by Industry and Type." Deloitte. https://www.deloitte.com/us/en/services/consulting/content/gen-ai-use-cases.html. Accessed August 24, 2025.

91. "AI Use Cases." EY. https://www.ey.com/en_us/services/ai/use-cases. Accessed August 24, 2025.
92. "601 Real-World Gen AI Use Cases from the World's Leading Organizations." Google. https://cloud.google.com/transform/101-real-world-generative-ai-use-cases-from-industry-leaders. Accessed August 24, 2025.
93. "The Most Valuable AI Use Cases for Business." IBM. https://www.ibm.com/think/topics/artificial-intelligence-business-use-cases. Accessed August 24, 2025.
94. "Artificial Intelligence (AI) in Manufacturing." Intel. https://www.intel.com/content/www/us/en/learn/ai-in-manufacturing.html. Accessed August 24, 2025.
95. "Beyond the Hype: Capturing the Potential of AI and Gen AI in Tech, Media, and Telecom." McKinsey & Company. https://www.mckinsey.com/~/media/mckinsey/industries/technology%20media%20and%20telecommunications/high%20tech/our%20insights/beyond%20the%20hype%20capturing%20the%20potential%20of%20ai%20and%20gen%20ai%20in%20tmt/beyond-the-hype-capturing-the-potential-of-ai-and-gen-ai-in-tmt.pdf. Accessed August 24, 2025.
96. "AI-Powered Success—With More Than 1,000 Stories of Customer Transformation and Innovation." Microsoft. https://www.microsoft.com/en-us/microsoft-cloud/blog/2025/07/24/ai-powered-success-with-1000-stories-of-customer-transformation-and-innovation/. Accessed August 24, 2025.
97. "Generative AI and Its Use Cases for Enterprise Applications." Oracle. https://blogs.oracle.com/ai-and-datascience/post/generative-ai-and-its-usecases-for-enterprise-apps. Accessed August 24, 2025.
98. "AI-Driven Revolution: How to Drive Your Business Success with Strategic AI Use Cases." PWC. https://pages.pwc.de/applied-ai-compass. Accessed August 24, 2025.
99. "SAP Business AI Use Cases." SAP. https://www.sap.com/products/artificial-intelligence/use-cases.html. Accessed August 24, 2025.
100. "Navigating the Data Minefields: Management's Guide to Better Decision-Making." Routledge. https://www.routledge.com/Navigating-the-Data-Minefields-Managements-Guide-to-Better-Decision-Making/Shemwell/p/book/9781032677934#top. Accessed April 7, 2025.
101. "Thinking Machines: The Search for Artificial Intelligence." Science History Institute. https://www.sciencehistory.org/stories/magazine/thinking-machines-the-search-for-artificial-intelligence/?msclkid=8d393fd2940e14a81080582632047153&utm_source=bing&utm_medium=cpc&utm_campaign=SHI%20-%20Algorithm&utm_term=sciencehistory&utm_content=Group%202. Accessed April 7, 2025.
102. "History of Artificial Intelligence (AI)." Britannica. https://www.britannica.com/science/history-of-artificial-intelligence. Accessed April 7, 2025.
103. "Can Machines Think?" The Rapid Response Institute. https://therrinstitute.com/can-machines-think/. Accessed April 7, 2025.
104. "Can Machines Think?" The Rapid Response Institute. https://therrinstitute.com/can-machines-think/. Accessed April 7, 2025.
105. "Artificial Intelligence Awarded Two Nobel Prizes for Innovations That Will Shape the Future of Medicine." Digital Medicine. https://www.nature.com/articles/s41746-024-01345-9. Accessed April 7, 2025.
106. "How CAD Has Evolved Since 1982." Scan2CAD. https://www.scan2cad.com/blog/cad/cad-evolved-since-1982/. Accessed April 5, 2025.
107. "Prometheus." Wikipedia. https://en.wikipedia.org/wiki/Prometheus. Accessed April 5, 2025.
108. "Frankenstein; or, The Modern Prometheus." Britannica. https://www.britannica.com/topic/Frankenstein-or-The-Modern-Prometheus. Accessed April 5, 2025.
109. "Sun Tzu versus AI: Why Artificial Intelligence Can Fail in Great Power Conflict." U.S. Naval Institute. https://www.usni.org/magazines/proceedings/2021/may/sun-tzu-versus-ai-why-artificial-intelligence-can-fail-great-power. Accessed April 5, 2025.
110. "David Eagleman: Neuroscientist, Author, Technologist, Entrepreneur." David Eagleman. https://eagleman.com/. Accessed April 6, 2025.

111. "EP98 'What's the Future of AI Relationships?'" iHeart. https://www.iheart.com/podcast/1119-inner-cosmos-with-david-e-110885566/episode/ep98-whats-the-future-of-ai-271387144/?keyid%5B0%5D=Inner%20Cosmos%20with%20David%20Eagleman&keyid%5B1%5D=EP98%20%22What%27s%20the%20future%20of%20AI%20relationships%3F%22%20%28with%20Bethanie%20Maples%29&sc=podcast_widget. Accessed April 6, 2025.
112. "People Are Falling in Love with AI. Should We Worry?" LIVESCIENCE. https://www.livescience.com/health/relationships/people-are-falling-in-love-with-ai-should-we-worry. Accessed April 6, 2025.
113. "Online Communication and Adolescent Well-Being: Testing the Stimulation versus the Displacement Hypothesis." Journal of Computer-Mediated Communication. https://academic.oup.com/jcmc/article/12/4/1169/4582968. Accessed April 6, 2025.
114. "Generative AI and the Future of Work in America." McKinsey Global Institute. https://www.mckinsey.com/mgi/our-research/generative-ai-and-the-future-of-work-in-america. Accessed April 5, 2025.
115. "Generative AI and the Future of Work in America." McKinsey Global Institute. https://www.mckinsey.com/mgi/our-research/generative-ai-and-the-future-of-work-in-america. Accessed April 6, 2025.
116. "AI or AU?" The Rapid Response Institute. https://therrinstitute.com/ai-or-au/. Accessed April 6, 2025.
117. "Navigating the Data Minefields: Management's Guide to Better Decision-Making." Routledge. https://www.routledge.com/Navigating-the-Data-Minefields-Managements-Guide-to-Better-Decision-Making/Shemwell/p/book/9781032677934#top. Accessed April 6, 2025.
118. "James Reason's 12 Principles of Error Management." Aerossurance. https://aerossurance.com/safety-management/james-reasons-12-principles-error-management/. Accessed October 9, 2025.
119. "Louis Pasteur: 'Chance Favors the Prepared Mind.'" The Socratic Method. https://www.socratic-method.com/quote-meanings-and-interpretations/louis-pasteur-chance-favors-the-prepared-mind. Accessed April 6, 2025.
120. "How I Realized AI Was Making Me Stupid—And What I Do Now." The Wall Street Journal. https://www.wsj.com/tech/ai/how-i-realized-ai-was-making-me-stupidand-what-i-do-now-5862ac4d. Accessed April 6, 2025.
121. "Voltaire: 'If God did not exist, it would be necessary to invent Him.'" The Socratic Method. https://www.socratic-method.com/quote-meanings-french/voltaire-if-god-did-not-exist-it-would-be-necessary-to-invent-him. Accessed April 5, 2025.
122. "Navigating the Data Minefields: Management's Guide to Better Decision-Making." Routledge. https://www.routledge.com/Navigating-the-Data-Minefields-Managements-Guide-to-Better-Decision-Making/Shemwell/p/book/9781032677934#top. Accessed April 6, 2025.
123. "Agentic AI." PEGA. https://www.pega.com/agentic-ai?utm_source=bing&utm_medium=cpc&utm_campaign=B_US_NonBrand_AgenticAI_CE_Exact_(CPN-108049)_EN&utm_term=agentic%20ai&gloc=73290&utm_content=pcridllpkwlkwd-80951896696119:loc-4126lpmtlelpdvlcl&gclid=40982402afc31319cb0f4c167cf8a1e3&gclsrc=3p.ds&msclkid=40982402afc31319cb0f4c167cf8a1e3. Accessed May 2, 2025.
124. "Agentic AI Explained: From Basic Concepts to Autonomous Decision-Making (FAQs)." AI GPT JOURNAL. https://aigptjournal.com/ai-resources/faqs/agentic-ai-explained/. May 2, 2025.
125. "What Are the Latest AI Developments Oil and Gas Needs to Be Aware Of?" Rigzone. https://www.rigzone.com/news/what_are_the_latest_ai_developments_oil_and_gas_needs_to_be_aware_of-28-apr-2025-180359-article/?utm_campaign=WEEKLY_2025_05_02&utm_source=GLOBAL_ENG&utm_medium=EM_NW_F1. Accessed May 2, 2025.
126. "AI and Expert System Myths, Legends, and Facts." IEEE EXPERT. https://www.ri.cmu.edu/pub_files/pub4/fox_mark_s_1990_2/fox_mark_s_1990_2.pdf. Accessed April 8, 2025.

127. "AI and Expert System Myths, Legends, and Facts." IEEE EXPERT. https://www.ri.cmu. edu/pub_files/pub4/fox_mark_s_1990_2/fox_mark_s_1990_2.pdf. Accessed April 8, 2025.
128. "Leadership: Pattonesque Style." The Rapid Response Institute. https://therrinstitute. com/leadership-pattonesque-style/. Accessed July 27, 2025.
129. "Pen and Culture: The Intricate Dance of Writing and Cultural Influence." The Science of Writing. https://scienceofwriting.org/pen-and-culture-the-intricate-dance-of-writing- and-cultural-influence/#:~:text=Writing%20is%20more%20than%20just%20a%20 tool%20for,diversity%2C%20setting%20norms%2C%20and%20establishing%20 rich%20writing%20traditions. Accessed June 17, 2025.
130. "The Theory of Language." Integrational Linguistics. https://www.integrational- linguistics.science/inteling/introTheoryLang. Access June 17, 2025.
131. "Biblical Bombshell: Mysterious Dead Sea Scrolls Decrypted with AI to Reveal Accu- rate Date." Honest AI. https://honestaiengine.com/biblical-bombshell-mysterious-dead- sea-scrolls-decrypted-with-ai-to-reveal-accurate-date?utm_source=ainews.honestaiengine. com&utm_medium=newsletter&utm_campaign=ai-hacks-that-ll-skyrocket-your- rankings&_bhlid=713024a41cd1c1f88df19ec5663622897a035ce7. Accessed June 17, 2025.
132. "Aeneas Transforms How Historians Connect the Past." Google DeepMind. https:// deepmind.google/discover/blog/aeneas-transforms-how-historians-connect-the-past/? utm_source=Generative_AI&utm_medium=Newsletter&utm_campaign=google- deepmind-decodes-the-past-by-training-on-ancient-texts&_bhlid=f49b5a4cf57b 77554882f5abd47fb30ccff7d722. Accessed July 24, 2025.
133. "Restoring and Attributing Ancient Texts Using Deep Neural Networks." Ithaca. https:// predictingthepast.com/ithaca. Accessed July 24, 2025.
134. "Vibe Coding: What Is It, And Where Is It Headed?" Forbes. https://www.forbes.com/ councils/forbesbusinesscouncil/2025/06/04/vibe-coding-what-is-it-and-where-is-it- headed/. Accessed October 3, 2025.
135. "AI Fabric: The Future of Data and AI Architecture." Fast Company. https://www. fastcompany.com/91284182/ai-fabric-the-future-of-data-and-ai-architecture. Accessed October 4, 2025.
136. "There's a Stunning Financial Problem with AI Data Centers." Futurism. https://futurism. com/data-centers-financial-bubble. Accessed October 4, 2025.
137. "Energy and AI." IEA. https://iea.blob.core.windows.net/assets/dd7c2387-2f60-4b60- 8c5f-6563b6aa1e4c/EnergyandAI.pdf. Accessed October 4, 2025.
138. "Massive Waste of Time." The Rapid Response Institute. https://therrinstitute.com/ massive-waste-of-time/. Accessed July 26, 2025.
139. "Unlocking the AI-First Organization: An Agentic Shift." Boston Consulting Group. https://media-publications.bcg.com/AI-First-Organization.pdf. Accessed July 26, 2025.
140. "Evolutionary Optimization of Model Merging Recipes." Nature. https://www.nature. com/articles/s42256-024-00975-8. Accessed July 27, 2025.
141. https://huggingface.co/spaces/open-llm-leaderboard/open_llm leaderboard#/. Accessed July 27, 2025.
142. "What Is ADAS (Advanced Driver Assistance Systems)?" synopsys. https://www. synopsys.com/glossary/what-is-adas.html#2. Accessed October 4, 2025.
143. "Corporate Average Fuel Economy." NHTTSA. https://www.nhtsa.gov/laws-regulations/ corporate-average-fuel-economy. Accessed October 4, 2025.
144. "Is Nuclear Power Set for a Revival?" Goldman Sachs. https://www.goldmansachs. com/insights/articles/is-nuclear-power-set-for-a-revival?chl=em&plt=briefings&cid= 0613&plc=body. Accessed October 4, 2025.
145. "Urania." Greek Gods & Goddesses. https://greekgodsandgoddesses.net/goddesses/ urania/. Accessed April 20, 2025.
146. "AI 'Urania' Transforms Gravitational Wave Detection with Revolutionary Designs." OPENTOOLS. https://opentools.ai/news/ai-urania-transforms-gravitational-wave- detection-with-revolutionary-designs. Accessed April 20, 2025.

147. "All about the Einstein Field Equations." Physics Forums. https://www.physicsforums.com/insights/all-about-the-einstein-field-equations/. Accessed April 20, 2025.
148. "AI 'Urania' Transforms Gravitational Wave Detection with Revolutionary Designs." OPENTOOLS. https://opentools.ai/news/ai-urania-transforms-gravitational-wave-detection-with-revolutionary-designs#section19. Accessed April 21, 2025.
149. "AI Discovers Innovative Detector Designs, Revolutionizing Gravitational Wave Detection and Human-Machine Collaboration in Science." Quantum Zeitgeist. https://quantumzeitgeist.com/ai-discovers-innovative-detector-designs-revolutionizing-gravitational-wave-detection-and-human-machine-collaboration-in-science/. Accessed April 20, 2025.
150. "When Machines Dream: AI Designs Strange New Tools to Listen to the Cosmos." SciTechDaily. https://scitechdaily.com/when-machines-dream-ai-designs-strange-new-tools-to-listen-to-the-cosmos/. Accessed April 21, 2025.
151. "How to Choose the Right AI Model for Your Application?" LeewayHertz. https://www.leewayhertz.com/how-to-choose-an-ai-model/. Accessed October 4, 2025.
152. "The Role of AI in Cybersecurity." CrowdStrike. https://www.crowdstrike.com/en-us/cybersecurity-101/artificial-intelligence/. Accessed October 4, 2025.
153. "Global Financial Institutions and Technology Leaders Collaborate under FINOS to Launch Open Source Common Controls for AI Service." FINOS. https://www.finos.org/press/global-financial-institutions-and-technology-leaders-collaborate-under-finos-to-launch-open-source-common-controls-for-ai-services. Accessed October 2, 2025.
154. "Sam's Club Phasing Out Checkouts, Betting Big on AI Shopping." Honest AI. https://honestaiengine.com/sams-club-phasing-out-checkouts-betting-big-on-ai-shopping?utm_source=ainews.honestaiengine.com&utm_medium=newsletter&utm_campaign=sam-s-club-s-ai-checkout-hack-shopping-just-got-a-futuristic-makeover&_bhlid=5187c494e649b46786017ea61cdd0293c4acfd99. Accessed April 21, 2025.
155. "Why ChatGPT's Essays Don't Fool the Experts – Yet." SciTechDaily. https://scitechdaily.com/why-chatgpts-essays-dont-fool-the-experts-yet/. Accessed May 28, 2025.
156. "ChatGPT Can Write Like You – Here's How You Can Get AI to Match Your Style and Tone." Honest AI. https://www.bing.com/search?q=statistical+bias+types&gs_lcrp=EgRlZGdlKgcIARBFGMIDMgcIABBFGMIDMgcIARBFGMIDMgcIAhBFGMIDMgcIAxBFGMIDMgcIBBBFGMIDMgcIBRBFGMIDMgcIBhBFGMIDMgcIBxBFGMID0gELNTgxNzYxNGowajSoAgiwAgE&FORM=ANAB01&PC=HCTS. Accessed May 28, 2025.
157. "Advanced AI Detector and AI Checker for ChatGPT & More." Undetectable AI. https://undetectable.ai/?utm_source=bing&utm_medium=cpc&utm_campaign=search&h_campaign_id=569461598&bng_id=1173180732210785&h_ad_id=73324000916162&msclkid=4078527d921d10e0d0b43a4a1571c408. Accessed May 28, 2025
158. "Introduction to Torque." LibreTexts-PHYSICS. https://phys.libretexts.org/Courses/Prince_Georges_Community_College/General_Physics_I%3A_Classical_Mechanics/38%3A_Torque/38.01%3A_Introduction_to_Torque. Accessed February 24, 2025.
159. "Truly Autonomous AI Is on the Horizon." UTS. https://www.uts.edu.au/news/tech-design/truly-autonomous-ai-horizon. Accessed February 24, 2025.
160. "Scientists Unveil AI That Learns without Human Labels – A Major Leap toward True Intelligence!" SciTechDaily. https://scitechdaily.com/scientists-unveil-ai-that-learns-without-human-labels-a-major-leap-toward-true-intelligence/#:~:text=Researchers%20have%20introduced%20Torque%20Clustering%2C%20an%20AI%20algorithm,human-labeled%20data%2C%20making%20it%20more%20scalable%20and%20efficient. Accessed April 12, 2025.
161. "AAAI 2025 Presidential Panel on the Future of AI Research." Association for the Advancement of Artificial Intelligence. https://media.licdn.com/dms/document/media/v2/D4E1FAQFYpFzqA5J9Dw/feedshare-document-pdf-analyzed/B4EZYaVKfoHgAc-/0/1744198526072?e=1745452800&v=beta&t=v2TUMZuhHgKFY4-T4JEXUi1R6ZcwOREywlxjYcoON44. Accessed April 19, 2025.

162. "Book Summary: 'The Rise and Fall of the Great Powers' by Paul Kennedy." The Ratchet of Technology. https://techratchet.com/2020/04/25/book-summary-the-rise-and-fall-of-the-great-powers-by-paul-kennedy/. Accessed April 10, 2025.

163. "For Geopolitics, What AI Can't Do Will Be as Important as What It Can." Rand. https://www.rand.org/pubs/commentary/2025/04/for-geopolitics-what-ai-cant-do-will-be-as-important.html. Accessed April 10, 2025.

164. "Origin and History of *Science*." etymonline. https://www.etymonline.com/word/science. Accessed June 19, 2025.

165. "Early COVID-19 Research Is Riddled with Poor Methods and Low-Quality Results – A Problem for Science the Pandemic Worsened but Didn't Create." The Conversation. https://theconversation.com/early-covid-19-research-is-riddled-with-poor-methods-and-low-quality-results-a-problem-for-science-the-pandemic-worsened-but-didnt-create-220635. Accessed October 7, 2025.

166. "Technology Assessment in the Era of Minimum Viable Product (MVP)." The Rapid Response Institute. https://therrinstitute.com/technology-assessment-in-the-era-of-minimum-viable-product-mvp/. Accessed October 7, 2025.

167. "Early COVID-19 Research Is Riddled with Poor Methods and Low-Quality Results – A Problem for Science the Pandemic Worsened but Didn't Create." The Conversation. https://theconversation.com/early-covid-19-research-is-riddled-with-poor-methods-and-low-quality-results-a-problem-for-science-the-pandemic-worsened-but-didnt-create-220635. Accessed October 7, 2025.

168. "How Is Enthalpy Not Directly Measurable?" PhysicsForums. https://www.physicsforums.com/threads/how-is-enthelpy-not-directily-measurable.808479/. Accessed October 7, 2025.

169. "Intelligence without a Brain." John Templeton Foundation. https://www.templeton.org/news/intelligence-without-a-brain. Accessed April 20, 2025.

170. "Who Was William James and How Did He Influence the Field of Psychology?" SimplyPsychology. https://www.simplypsychology.org/william-james.html. Accessed April 21, 2025.

171. "Teleology." Ethics Explainer. https://ethics.org.au/ethics-explainer-teleology/. Accessed April 21, 2025.

172. "Technological Approach to Mind Everywhere: An Experimentally-Grounded Framework for Understanding Diverse Bodies and Minds." National Library of Medicine. https://pmc.ncbi.nlm.nih.gov/articles/PMC8988303/. Accessed April 21, 2025.

173. "Technological Approach to Mind Everywhere: An Experimentally-Grounded Framework for Understanding Diverse Bodies and Minds." National Library of Medicine. https://pmc.ncbi.nlm.nih.gov/articles/PMC8988303/. Accessed April 21, 2025.

174. "Motor Behavior Is an Example of an Unconventional Intelligence." Benjamin Lyons. https://interestingessays.substack.com/p/motor-behavior-is-an-example-of-an?utm_source=substack&utm_medium=email. Accessed April 21, 2025.

175. "Muscle Memory." Cleveland Clinic. https://my.clevelandclinic.org/health/articles/muscle-memory. Accessed April 21, 2025.

176. "What Is Distributed Computing?" AWS. https://aws.amazon.com/what-is/distributed-computing/. Accessed April 21, 2025.

177. "What Is Latent Variable?" geeksforgeeks. https://www.geeksforgeeks.org/what-is-latent-variable/. Accessed February 27, 2025.

178. "Analysis of Variables That Are Not Directly Observable: Influence on Decision-Making during the Research Process." SciFlo Brazil. https://www.scielo.br/j/reeusp/a/nNB3rfKQ6xJ6Bq6KTSV6VvK/. Accessed March 7, 2025.

179. "Analysis of Variables That Are Not Directly Observable: Influence on Decision-Making during the Research Process." SciFlo Brazil. https://www.scielo.br/j/reeusp/a/nNB3rfKQ6xJ6Bq6KTSV6VvK/?format=html&lang=en. Accessed October 7, 2025.

180. "Behavioral Science." Merriam-Webster. https://www.merriam-webster.com/dictionary/behavioral science. Accessed March 6, 2025.

181. "Game Theory." Stanford Encyclopedia of Philosophy. https://plato.stanford.edu/entries/game-theory/. Accessed March 6, 2025.
182. "What Is Systems Dynamics?" System Dynamics Society. https://systemdynamics.org/what-is-system-dynamics/. Accessed March 6, 2025.
183. "Observer Bias: Definition, Examples & Prevention." SimplyPsychology. https://www.simplypsychology.org/observer-bias-definition-examples-prevention.html. Accessed March 6, 2025.
184. "Physics." Britannica. https://www.britannica.com/science/physics-science. Accessed February 27, 2025.
185. "The Importance of Behavioral Science." mindworks. https://www.chicagobooth.edu/mindworks/what-is-behavioral-science-research. Accessed February 27, 2025.
186. "The Fascinating World of Behavioral Sciences: Unraveling the Mysteries of Human Behavior." Journal of Public Health and Nutrition. https://www.alliedacademies.org/articles/the-fascinating-world-of-behavioral-sciences-unraveling-the-mysteries-of-human-behavior-26957.html. Accessed February 27, 2025.
187. "The Sveriges Riksbank Prize in Economic Sciences in Memory of Alfred Nobel 1994." The Royal Swedish Academy of Sciences. https://www.nobelprize.org/prizes/economic-sciences/1994/press-release/. Accessed March 7, 2025.
188. "Game Theory Review: Key Term – Non-Cooperative Game." Fiveable. https://library.fiveable.me/key-terms/game-theory/non-cooperative-game. Accessed March 8, 2025.
189. Mandel, Michael J. (1994, October 24). "How Game Theory Rewrote All the Rules." Business Week, p. 44.
190. "Notes on Non-Cooperative Game Theory." Iñaki Aguirre. https://www.ehu.eus/iaguirre/MicroIVEnglish/NotesGames2009.pdf. Accessed March 8, 2025.
191. "How IBM's Deep Blue Beat World Champion Chess Player Garry Kasparov." IEEE. https://spectrum.ieee.org/how-ibms-deep-blue-beat-world-champion-chess-player-garry-kasparov. Accessed October 7, 2025.
192. "Strategic Intelligence in Large Language Models: Evidence from Evolutionary Game Theory." arXiv.org. https://arxiv.org/pdf/2507.02618. Accessed October 7, 2025.
193. "Navigating the Data Minefields: Management's Guide to Better Decision-Making." Scott M. Shemwell. https://www.routledge.com/Navigating-the-Data-Minefields-Managements-Guide-to-Better-Decision-Making/Shemwell/p/book/9781032677934#top. Accessed March 8, 2025.
194. "Cross-Cultural Negotiations between Japanese and American Businessmen: A Systems Analysis (Exploratory Study)." Scott Shemwell. https://www.proquest.com/openview/ac7877868846620edd6444d031e306bf/1?pq-origsite=gscholar&cbl=18750&diss=y. Accessed February 27, 2025.
195. "Structural Dynamics: Foundation of Next Generation Management Science (Changing the Dialogue)." The Rapid Response Institute. https://www.amazon.com/Structural-Dynamics-Foundation-Generation-Management-ebook/dp/B00U0JKMT0?ref_=ast_author_dp&dib=eyJ2IjoiMSJ9.ZsgjVH3gE-2bNlG58UFbovkVn3k95ZDcAcQNTKiMzOxED9JOD_W-LRGQqyztnLI-n453mVW8QYOE6akjfGEFE-sED4H3bL7PJcnKFk3yKZc.2efLOtwXeXqioMRyLyPbmOPDP10cDkwTlse0e9ytXCw&dib_tag=AUTHOR. Accessed October 8, 2025.
196. "All the Pundit Were Wrong—Again!" Governing Energy. https://therrinstitute.com/wp-content/uploads/2018/10/All-the-Pundits-Were-Wrong%E2%80%94Again-November-17-2016.pdf. Accessed October 7, 2025.
197. "Structural Equation Modeling with Many Variables: A Systematic Review of Issues and Developments." Frontiers in Psychology. https://www.frontiersin.org/journals/psychology/articles/10.3389/fpsyg.2018.00580/full. Accessed October 7, 2025.
198. "What Is the Model Context Protocol (MCP)?" Model Context Protocol. https://modelcontextprotocol.io/docs/getting-started/intro. Accessed October 7, 2025.
199. "What Is Decentralized AI Model." Geeksforgeeks. https://www.geeksforgeeks.org/blogs/what-is-decentralized-ai-model/. Accessed October 7, 2025.

2 Why Did the Data Deceive Me?

Data is like garbage. You'd better know what you are going to do with it before you collect it.

—**Mark Twain**[1]

We humans can infer things where none exist. Not imagination or fantasy, but the assessment of data or partial data sets where the end decision product is not just incorrect, but even implausible. When a more robust and objective review is undertaken by experts with minimal bias, some issues become clearer, such as the US presidential elections (Section 4.3.2.1.1 'US Presidential Election Survey' of Chapter 4).

We address issues of data quality and management in detail in our 2025 book, *Navigating the Data Minefields: Management's Guide to Better Decision-Making.*[2] Here we expand on that information with a few examples where the data was not what humans interpreted to be.

2.1 THE MAGDEBURG UNICORN

Data issues have always plagued humankind. One interesting story is the Magdeburg Unicorn. In 1663, scientists found ancient bones in Germany that were interesting jumble of remnants. Piecing the bones together resulted in the skeleton of a unicorn. However, what they had incorrectly assembled was a collection of different bones. The mythical creature still has lessons for those responsible for decisions based on data. These include[3]:

- **The Misalignment of Data Sources**—Most organizations base decisions using data from various sources, legacies, quality, and gaps. The utmost care must be taken by data scientists, subject matter experts, and others to assure the normalization of structured data and quality and integrity of unstructured data. Abnormalities must be addressed before going forward with any analysis.
- **Fragmented Information**—Gaps or missing data/information should be assessed as their importance and potential impact this may on AI solutions. Appropriate remedies must be taken or at least the risk understood at an acceptable level.
- **Reliance on Legacy Data**—This can be a very difficult task for two reasons. First, by definition the data is old, and its collection process may be poorly documented/understood. In addition, depending on the type of data, for example, real-time, the sensor/collection processes for new data may be

DOI: 10.1201/9781003646914-2

different and more robust. All data sets used in an analysis will need to be reconciled.

Data Reconciliation is "a systematic process of comparing, verifying, and ensuring data consistency and accuracy across different datasets and systems."[4] It is a major technical project that requires expertise in this area. Management should consult with subject matter experts in this field.

- **Overly Creative Analysis**—The data cannot give something that is not there. Imagination and desires must be kept in check, especially for complex analysis involving tens of millions or more data points.
- **Ignoring Context**—Assure that team members consist of knowledgeable individuals with tools to assure data is not taken out of context.
- **Poor Communication and Collaboration**—This is a continual problem for project teams as well as their executive sponsor and end user client(s). This issue is addressed further in Section 5.4.2 'Key AI Developer Skills Necessary' of Chapter 5.
- **Independent Validation and Verification (IV&V)**—This comprehensive review by an independent third-party skilled in this process should be mandatory for any capital AI project.

All of these potential issues listed should be addressed in the organization's data governance policies and procedures. Organizations can use an Operations Management System (OMS) to assure these processes are adhered to (Section 3.2.1 'Operations Management System' of Chapter 3).

Also known as the Magdeburgian Einhorn or Guericke-Einhorn, this mythological creature received over 200,000 likes on Twitter and Reddit. A comical representation of failed data management but scientists of the time must have taken it seriously.[5] This appears to also be a case of bias in that clearly some believed unicorns existed and that belief system factored into the logic and decision-making process.

2.2   LEGACY DATA

This spreadsheet approach was shown in the recent US elections to be detrimental to the party mainly depending on this view of society. How many times did we hear pollsters opine that X percent of a certain population was voting for one candidate vs. the other?

The categories of people were divided along traditional lines. This whole model was shattered on November 5, 2024, and likely some providers of data in this format may no longer be in business for the next cycle.

However, this belief system is not limited to the political class. We all can fall into this stereotype.

For decades we have been taught that data categories can be captured under a Normal Distribution (Bell) Curve of a *category* and *row* of interest, i.e., the distribution of the height of a class of male senior high school students or SAT scores. Another more relevant example, retailers' pertinacious obsession with the 18–34-year-old group.

These are fairly simple models, and in this Blogger's opinion, this representation rarely works anymore, if it ever did. For example, the resulting inventory overages as a result of dependencies on this gross data model are not an efficient return on shareholder value. Other recent missteps based on faulty interpretation of the customer/prospect base include Bud Lite advertising fiasco and the Target marketing failure.

It is ok to make marketing mistakes, that is going to happen. The problem with these two (and other) campaigns is the analysis of risk, return, etc. was likely shallow or mathematically primitive.

There are ways to appeal to new consumers without alienating a large existing base. It's all in big numbers.

There are several validated models of human behavior, and we discuss two of them herein, Maslow's Hierarchy and the Relationships, Behaviors, Conditions model. Some readers may prefer others, and they will most likely work within this construct as well.

Finally, Monday Morning Quarterbacks argue that their polls were accurate. Many are often those who conduct internal sampling for the candidates. Regardless, the public was told by the so-called *reputable and seemingly knowledgeable sources* that the presidential race was tight and a close 'horse race.'

In fact, it was not and off by a wide margin. Unless a strong bias or worse was at work with these analyses and reporting, something is still amiss!

2.3 RELATIONSHIPS, BEHAVIORS, AND CONDITIONS (RBC)

This pundit has been an advocate of the RBC model since first discovered in the early 1990s. This model is robust and captures the essence of complex human relationships in a straightforward manner. We developed this model in detail in *Navigating the Data Minefields: Management's Guide to Better Decision-Making* and interested readers are invited to check it out. Therefore, we briefly touch the high points here as we refer this model routinely in our AI thought process.

The following is taken from our Cross-Cultural Online Game.[6]

The **RELATIONSHIPS**, **BEHAVIORS**, and **CONDITIONS** (RBC) model was originally developed to address issues around cross cultural (international) negotiation processes. **Relationships** are the focal point of this perspective, reflecting commonality of interest, balance of power and trust as well as intensity of expressed conflict. **Behaviors** in this model are defined as a broad term including multi-dimensions and intentional as well as unintentional. Finally, **Conditions** are defined as active and including circumstances, capabilities and skills of the parties, culture, and the environment. Of course, **Time** is a variable in this model as well.

As shown in Figure 2.1, Relationships are directly affected by both Behaviors as well as Conditions. Note that Conditions also directly impact on Relationships as well.

Moreover, we have defined *Behavioral Economics* as "the decision-making model that incorporates societal, cultural, emotions and other human biases into the process as opposed to the classic rational economic actor."

One key feature of the R B C Framework is its emphasis on interactive relationships while providing an environment for multiple levels of behavioral analysis. This

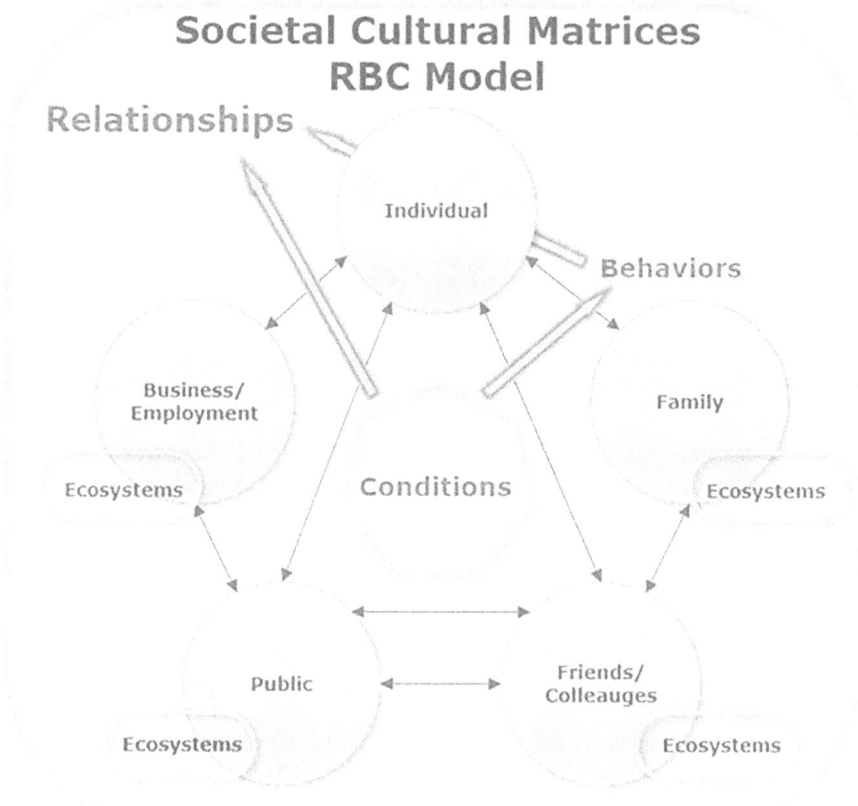

FIGURE 2.1   Societal cultural matrices—RBC model.

makes it a useful tool to better understand the new Big Data/AI processes currently unfolding.

Figure 2.1 is a derivative of our Cross-Cultural Interactions model. It is a peer-reviewed model and is a very good way to calibrate interactions.

Five Elements—As depicted in the graphic, organizational cross-cultural interactions incorporate one or more of these dynamic components.

- **Individual**—A single human being or perhaps other entity with whom others collaborate
- **Internal Organization**—The (legal) entity one or more individuals associate with on a daily basis
- **Government**—Elected officials as all levels of the Federal, State, and Local government regulatory bodies including any and all regulatory bodies or other agencies

- **Public**—Communities and other interest groups including the media as well as their social aspect
- **Ecosystem**—The broad group of constituents which includes clients/customers, the supply chain as well professional services engaged with the organization

Sub-cultures abound across the four elements other than the individual. For example, a taskforce or team may be formed to implement a global IT system. In this case, the team may be composed of individual from several countries as well as Internal and External IT professionals. In this case this would be an Internal Organization Sub-Culture. Another Sub-Culture might be the NYC office with a distinct local flavor.

Finally, the RBC model is overlaid over the five societal cultural elements. We believe this the social framework that AI systems dealing with human behaviors must use in behavioral decision-making processes.

This author has long believed that we do not live in a linear or mathematically deterministic world. A strong belief in stochastic, matrices drives much of my thinking. This is reflected in the figure on the left. The RBC model is the foundation for the five parties and their collective Behaviors. The resulting Relationships range from one-on-one to small groups or in some cases a wide and varied constituency.

Note that there is an Ecosystem (*something, such as a network of businesses, considered to resemble an ecological ecosystem especially because of its complex interdependent parts*) associated with each entity other than the individuals, which are the other four.

This is a wide array of influences on any given individual. Not easily measured, linearly!

2.4   THE DATA MANAGEMENT PROBLEM

If we cannot determine whether a spreadsheet is valid and reliable, how can we be assured that AI calculations using perhaps tens of millions or more data points is an accurate solution to the problem prompted/put forth?

Data Management—As AI models become more sophisticated and to the extent a human cannot adequately determine its viability, very high data quality is the fail-safe position.

We have identified a general problem as well as three valid and reliable theories. This is a nice discussion but is without relevance with Valid, Reliable, and Timely (VRT) data.

Data Management has been a problem as long as there has been data, stone, paper, or electronic. It has always been an issue. "Herman Hollerith is given credit for adapting the punch cards used for weaving looms to act as the memory for a mechanical tabulating machine, in 1890." Much later (circa 1950s) digital database management schemas emerged. Today, we are flooded by data of all types including deliberately false data/information. This paragraph is taken from a well-done overview of the history of database management and well worth a quick read. It puts much of our current data discussions in context.[7]

How can this be effectively managed and how can an executive without a technology background survive much less thrive in such an environment?

For decades, this author has been involved in almost all computer system development from the yellow punch tape of the 1960s until today. There have been several constants over that period.

One of the biggest chronic challenges has been database management; from the acquisition of data through its life cycle until achieve. One of the outcomes of poor data management has been decisions that have even led to the demise of firms, along with tens of thousands (if not more) of suboptimal decisions at the expense of shareholder value (bottom line/stock price). What makes the current crop of Big Data/Artificial Intelligence advocates smarter than those who came before? Nothing! Unless some things are changed.

The following and other managerial action items are developed in detail *Navigating the Data Minefields: Management's Guide to Better Decision-Making*.

- How to determine the quality of the data and its relevance to the decision-making processes.
- Current Challenges and Trends in Big Data and associated applications.
- Proposed organization structure such as High Reliability Organization and an understanding of Human Factors to fully realize the full and measurable economic value from these technologies.
- A full set of Risk Mitigation and a Governance model including Disaster Recovery and Cyber Security.
- How these technologies are used in Operations, a proposed Management System as well as numerous Case Studies across a number of industries and types of problems.

The challenge of managing this suite of emerging AI/Big Data is daunting and one that cannot be dodged or delegated. How organizations respond can be the difference between success or the destruction of shareholder value.

2.5   MASLOW'S HIERARCHY OF NEEDS

When we started looking at the results of the 2024 US presidential election, we recognized that many of the data analysis issues used appeared to be inherently flawed. As part of that investigation, we took another look at an older model as well as our RBC framework.

As documented in referenced 2024 blog, the following approach emerged as perhaps a way of replacing the older polling models or at least augmenting them with new thinking. Moreover, this framework is perfect for AI, and the host of behavioral and other latent variables tradition sampling often miss.

One of the most influential psychologist of the 20th century, Abraham Maslow, made many contributions to this field. Perhaps his most famous (1943) legacy is his hierarchy of human needs. A revised edition of this five-step pyramid is depicted and defined in Figure 2.2. He logically puts forth a logical process that humans need basic needs, i.e., food, shelter, and safety before higher levels of humanity are of

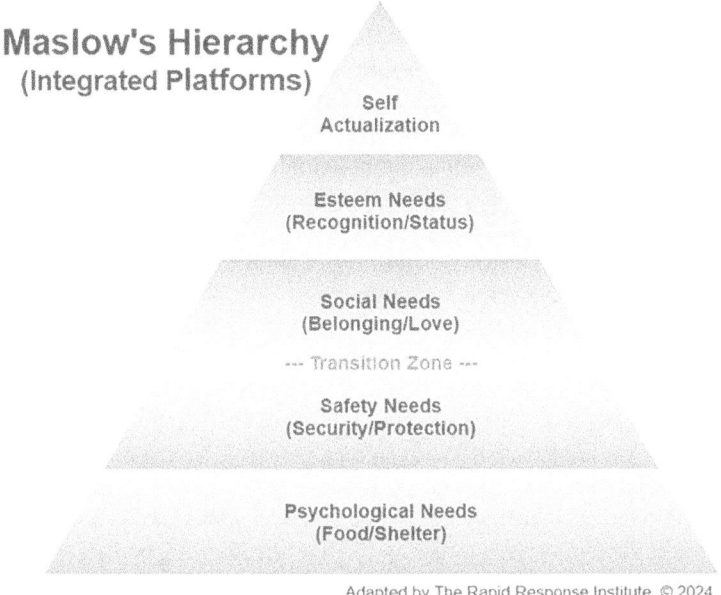

FIGURE 2.2 Maslow's integrated platform hierarchy.

interest. Logically, this is a working model that is useful for AI practitioners and the management responsible for organizations in this ultra-high-tech world.

In 1970 he made some important revisions to his initial theory, admitting that:

- The hierarchy's order is not as rigid as he initially proposed; instead, it is flexible based on external circumstances and individual differences.
- Most human behavior is multi-motivated, meaning that any behavior simultaneously aims to fulfill many needs.
- Three more levels could be added: cognitive needs, aesthetic needs, and transcendence needs (e.g., mystical, aesthetic, sexual experiences, etc.).

Cross-cultural research by Tay and Diener (2011) supports the view that there are universal human needs regardless of cultural differences, but the order of importance is influenced by culture."

Interesting, Maslow "lamented that the U.S. forces did not understand the German opposition and felt that the field of psychology could help facilitate understanding and restore peace to the world.[8]

We address a behavioral economics model that enables a better understanding between people and cultures. In Section 1.12.2.1 'Behavioral Science' of Chapter 1, we advanced this model to incorporate AI to increase the effectiveness of a model we put forth initially in the mid-1990s.

This perspective on human needs and subsequent behaviors dates back to the 1940s. As shown it consists of five steps from survival to the ultimate Self Actualization. Note that we adapt this model in two ways. First, we view this as a growth process as opposed to five discrete steps.

Additionally, the bottom two lower ranges focus mostly on the physical issues we all face. Fundamental survival and security both physically, mentally, emotionally, and spiritually. Once we have attained, sustained, and *believed*, we are ready to transition toward an end game, fulfilling our potentialities at our highest level.

The color gradient is meant to reflect that this journey is a process and often each field begins slowly before accelerating into the next range. Another perspective of each the Five Platforms consists of a number of smaller steps, which when assessed using Integral Calculus, generates a continuum.

2.5.1 CATEGORIES

As mentioned, we tend to clump or stereotype individuals into preconceived 'buckets'. Over time, these buckets have taken on an aura of, it's the only way. Basically, the construct that 'we have always done it this way.'

This is no longer a viable business or social model. Our world is much to dynamic for the older static approach to targeting the likelihood of a specific demographic to perform in a preconceived prescriptive manner.

One place to start looking is at the US Census data. This data set consists of 57 profiles, The United States and each State and Territory. Each Profile is divided into several Sub-Categories, Populations and People, Income and Poverty, Education, Employment, Housing, Health, Business and Economy, Race and Ethnicity, and Nearby States. Each Sub-Category is further detailed.

We might even call the data in these Categories/Sub-Categories, **Conditions** from the RBC model. Each **Platform** in Maslow's Hierarchy can be considered a Category with infinite Sub-Categories. Likewise, Conditions.

2.5.2 ROWS

We have identified the following three variable model as representative of human **Behaviors**—Wants, Needs, and Desires. We have previously reviewed some of this in our Blog.[9] These can also be mapped to Behaviors in both the RBC and Maslow models.

Details and a short list of some variables follow. These definitions can overlap, and readers will note that there is some duplication across all three classes below. Researchers use different definitions for each class. This does not detract from the fundamentals of Rows selected for analysis.

The point is to develop a list of behaviors that is indicative of the problem to be solved.

> **Wants**—For the purposes of this model, we define Want as something that an individual might seek as part of normal life, i.e., and ice cream cone.
> **Needs**—Those fundamentals of life, especially as defined in Maslow's Hierarchy, platforms one and two.

Desires—Different from Wants, Desires are perhaps beyond his/her reach. *Coveting* a promotion and willing to back stab co-workers to get it rather than compete on merit.

2.5.3 FINAL THOUGHTS

Data challenges remain and there is no silver bullet to fix these long-standing issues. Management must demand great attention to the data it uses to train and use AI. Failure to assure high quality data feeds will result in suboptimal or even erroneous results.

The devil as they say is in the data details. Again, in the AI era data is no longer just an IT issue and governance model must include internal and external data standards and expectations. This was addressed in detail in our previous book.

NOTES

1. "125 Inspirational Quotes About Data and Analytics [2025]." Digitaldefynd. https://digitaldefynd.com/IQ/inspirational-quotes-about-data-and-analytics/. Accessed June 11, 2025.
2. "Navigating the Data Minefields: Management's Guide to Better Decision-Making." Scott M. Shemwell. https://www.routledge.com/Navigating-the-Data-Minefields-Managements-Guide-to-Better-Decision-Making/Shemwell/p/book/9781032677934#top. Accessed June 11, 2025.
3. "Don't Build a Magdeburg Unicorn With Your Data." Jeff Winter. https://www.jeffwinterinsights.com/insights/magdeburg-unicorn. Accessed June 5, 2025.
4. "What is Data Reconciliation? Importance, Challenges, Use Cases and More." FinnOps.ai. https://www.finnops.ai/blogs/what-is-data-reconciliation. Accessed June 5, 2025.
5. "The Complicated History of the 'Magdeburg Unicorn'" Snopes. https://www.snopes.com/articles/441423/magdeburg-unicorn/. Accessed June 5, 2025.
6. "Cross Cultural Game." The Rapid Response institute. https://therrinstitute.com/lets-play-a-game/. Accessed October 7, 2025.
7. "A Brief History of Database Management." DATAVERSITY. https://www.dataversity.net/articles/brief-history-database-management/. Accessed October 7, 2025.
8. "Abraham Maslow, His Theory & Contribution to Psychology." PositivePsychology.com. https://positivepsychology.com/abraham-maslow/. Accessed June 23, 2025.
9. "Want – Like – Need." The Rapid Response Institute. https://therrinstitute.com/want-like-need/. Accessed October 7, 2025.

3 AI Implementation Framework

Fortunately, most human behavior is learned observationally through modeling from others.

—Albert Bandura[1]

It is beyond the scope of this book to address the major body of knowledge in project management. We believe that world class project management is also required for artificial intelligence (AI) transformations. As shown in this book and our previous work, there is a huge amount of detail that goes into such an effort.

In some ways, technology is the easy part. The human side is always the high-value component. We do not believe this will change even as AI tasks over menial tasks as well as high-performance medical research, safety and environment management, as well as the economy.

This chapter focuses on those items that are specific to AI. We do not expect this to be comprehensive because AI is evolving rapidly, and many changes are likely in the near future. Rather it is a framework that management can look for guidance.

3.1 SCIENTIFIC METHOD FOR AI

Most readers are familiar with how science has been brought into bolster the position of pundits/advocates of policy, especially related to COVID-19. The word is tossed around casually, as if most who use it know what they are talking about. Spoiler Alert–Most Do Not! This includes some with advanced technical or medical bona fides.

Most think that by using the word 'science' this assessment process is out of their wheelhouse. Previously, we argued that there is a layperson's approach that works fine for most of our daily needs. Details can be found on our Blog, They Blinded Me with Science.[2]

We can put the scientific method in lay terms and use common sense in our technology decision-making processes. Ask this short list of questions:

1. **Pose a Testable Question**—Ask yourself how can I measure the response?
2. **Conduct Background Research**—Google search and others, recognizing the probability of bias on the part of authors
3. **State your Hypothesis**—Question with NO pre-conceived outcome (Pseudo-Science)
4. **Design Experiment**—How can I test my hypothesis?
5. **Perform your Experiment**—Test your idea

DOI: 10.1201/9781003646914-3

6. **Collect Data**—Write down anything that you learn
7. **Draw Conclusions**—What makes logical sense (Mr. Spoke)?
8. **Publish Findings**—Tell your colleagues, write a blog or more

If your organization can meet the test of these eight points, it is effectively taking advantage of the scientific method. In any event, these points should be addressed early in the project/technology assessment process.

3.1.1 HUMAN INPUT STILL NEEDED

From our discussion above, pollsters and marketers will still need to ask questions that will also help structure the model, but the processes posited herein can change the weighting processes driving toward equilibrium and Pareto Optimality (Pareto efficiency implies that resources are allocated in the most economically efficient manner but does not imply equality or fairness) in the final analysis. In other words, higher quality results.

These numbers can also be seen as 'first value' in a simulation. Another way to look at this is to the logic the Turing Bombe (electromechanical code-breaking) Machine of World War II. Not setting a number as itself.[3]

Then AI can take the model to the next level and provide pollsters/marketers with real, modern solutions. Finally, other uses will most likely be derivatives of this solution.

Finally, at the current state of AI, critical thinking is required to assess the accuracy and relevance of outputs (Section 1.11.4.6 'Choosing the Right AI Model' of Chapter 1). Likely, future AI will still be a partner with human counterparts and overlords. We will be freed from computing drudgery to reach toward Maslow's Self-Actualization (Section 2.5 'Maslow's Hierarchy of Needs' of Chapter 2).

The human roles will change; however, replacement by AI will be tasks specific. New tasks will emerge as a result of AI and human skills will be needed (Section 4.2.9 'Role of Humans in the AI Era' of Chapter 4).

3.2 OPERATIONAL RISK MANAGEMENT

For most organizations, operations is the ability to generate profitable revenue. Therefore, it is an area where AI can generate significant value for organizations. As with any process of this magnitude there are elements of risks that must be recognized and mitigated.

An earlier version of the following graphic was published in our 2025 book, *Navigating the Data Minefields: Management's Guide to Better Decision-Making*. We have updated those risk mitigation processes to reflect the use of AI to better assess risks and develop mitigation strategies that minimize these exposures. The high-level description of the updated model is the following:

Three major ongoing AI risks:

- **Risk Identification and Assessment**—Rank order of exposure to a portfolio of possible risks

- **Risk Monitoring**—The process and technology systems necessary to monitor and provide alerts when out-of-limit events occur
- **Risk Measurement**—Incorporating risk monitoring into the Key Performance Indicators and Critical Success Factors (CSFs) used as the basis of decision-making and management.

The model is divided into several sections:

- **The Operational Risk Environment**—Focus on action items such as Vision, Guiding Principles, etc.
- **Ongoing Risk Management**—As described in Section 3.2 'Operational Risk Management' of Chapter 3.
- **General Aspects of the Business**—AI and data alignment required
- **The Maturity of the OT-IT Relationship**—The interrelationship between operations and information technology (IT), e.g., SCADA, Industrial Process Control (IPC).
- **Data Quality**—As defined in Chapter 2, Why Did the Data Deceive Me?
- **Outside/Third Party Stakeholders**—Those data users outside the firm's firewall
- **Comprehensive Risk Reporting**—Similar as developed by The Committee of Sponsoring Organizations of the Treadway Commission (COSO) and its Internal Control—Integrated Framework as well as in keeping with ISO 31000: Risk management—Guidelines

Note that best AI risks practices are not limited to the organization but must include its supply chain ecosystems (Figure 3.1). High-performance vendor and even customer management (ecosystem) is critical to success in this field and is further explored in Section 6.6.2.3 'Pareto Optimality' of Chapter 6. That use case documents processes that help assure risks are minimize.

3.2.1 OPERATIONS MANAGEMENT SYSTEM

An Operations Management System (OMS) is a collection of processes and procedures that enables a company to effectively manage business practices and achieve the highest level of Operational Excellence in daily operations.[4]

ISO/IEC 42001:2023 (see Appendix III) is an international standard that specifies requirements for establishing, implementing, maintaining, and continually improving an Artificial Intelligence Management System (AIMS) within organizations. It is designed for entities providing or utilizing AI-based products or services, ensuring responsible development and use of AI systems. Typically, we would integrate these guidelines into existing OMS models of which there are several, each designed to meet the needs of organizations in different industry sectors.

In Section 6.5 'Smart Manufacturing: The Re-Refining Process' of Chapter 6, we describe an AI-powered OMS that we have in conjunction with a client. As shown in Figure 3.2, AI is driving IT, and Operational Technology (OT), and may include Engineering Technology (ET) in the near future. This makes sense as AI looks

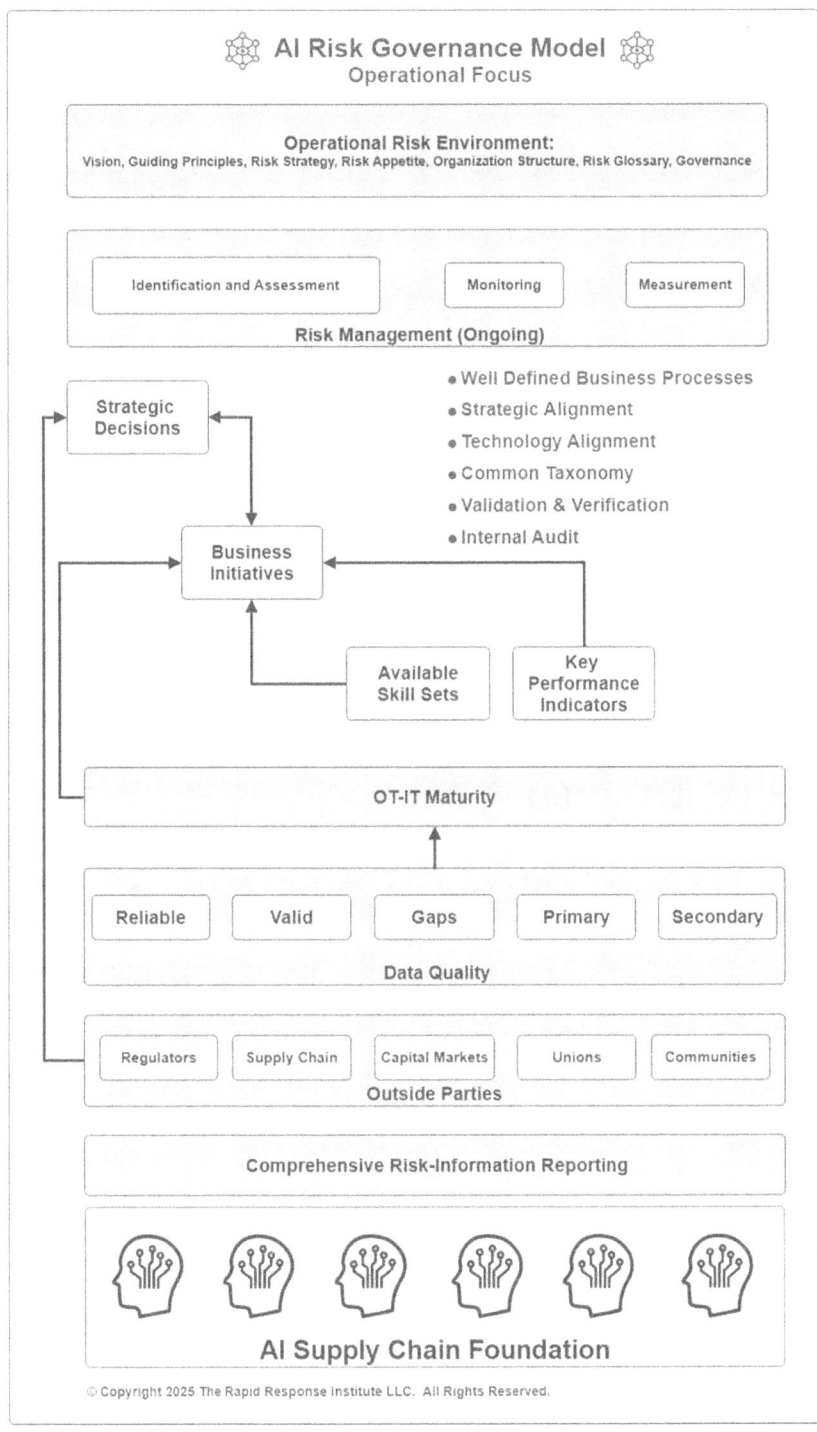

FIGURE 3.1 AI risk governance model: Operational focus.

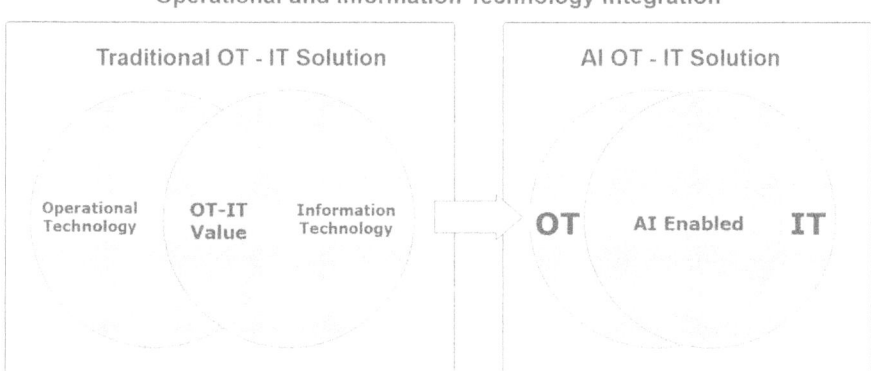

FIGURE 3.2 AI OT-IT value.

across organizational silos, hence that business model breaks and evolves toward a more integrated one.

As shown in Figure 3.2, we expect that AI will drive convergence and closer integration with most if not all organization departments. One example is the workflow-driven merge of Human Resources and IT.[5]

3.3 AI PROJECT METHODOLOGY

There are a number of AI Project Management techniques and a number of vendors offering solutions. Therefore, that decision is left to each organization and should be aligned with organizational needs, culture, and project management governance.

However, I always like to start with the Project Management Institute. In my opinion, this organization is the project management gold standard and a good starting point.

It can provide expertise leading AI Transformation (PMI Infinity (tm0) internally. Moreover, they offer a suite of AI solutions organizations can use to support other projects they may be involved with, i.e., new plant construction.[6]

In the remainder of this section, we address several key items to consider for AI implementation projects. As always, the list is not all inclusive and other CSFs may emerge.

3.3.1 AI FRAMEWORKS

AI frameworks offer significant advantages to the organization by standardizing AI development workflows. Their standard toolkit and methodology helps assure consistency and smooth integration across various platforms and applications. This is important for any technology and especially a new, immature one such as AI. They are a collection of libraries forming the core ingredients that improve the efficiency of algorithm development and deployment regardless of the develop team(s) composition, i.e., combination of internal and third-party developers.[7]

There are a number of AI frameworks available, both Open Source and Commercial. Deciding which one to use does not need to be a daunting task and you should follow good practices. Forming a multi-functional team lead by the CAIO is probably a good start. Applying best practice technology assessment processes should be the backbone.

Buy versus Build or combination, as in any technology or even mundane project is always the first decision. Most current documentation regarding AI appears to originate with the vendor community. They argue the value brought by third parties as well as the risk of an organizational Due It Yourself project. Likely your team will be comprised of internal individuals and outside experts.

Regardless, organizations must be knowledgeable buyers more than any IT project before. Even ERP solutions were built around process and regulatory, best practices, i.e., GAAP. It is organizational maleficence to turn AI development and implementation over to others without having the knowledge to assure the value of what is being built as well as its alignment with organizational values a strategic direction. After all, the end of the project, construction firms leave or take over long-term expensive maintenance and refresh programs.

While we know that the solution to the political pollster and/or marketing manager is not a definitive answer, we can do better than we are. We know how to manage data in this environment and we have models for assessing and making decisions based on results. As we move from simple, small data set, linear models to robust Big Data Analysis, we need to consider a few additional action items.

3.3.2 Problem Statement

Used improperly, technologies can and do result in the deterioration of cognitive faculties that ought to be preserved.[8]

As a child in the 1950s, I studied mathematics the old-fashioned way. Rote exercises with no understanding of the underlying theory gave me a level of analytical capabilities relevant for the time and my needs at that point in my life.

Later as a university student I learned how to use a slide rule, and scientific notation (expressing numbers that are too large/small depicted as decimals). I also learned how to set up complicated problems mathematically. This was necessary because without contemporary computing capabilities, we could not really solve the problem. We had to demonstrate we could properly frame the problem, hence an understanding or even knowledge.

3.3.2.1 Mathematics as a Language

In the 1950s, *New Math* was in vogue, although ultimately deemed a major failure. In 1951, the so-called father of new math, Max Beberman created a high school curriculum where math was treated more as a language than science. "He encouraged students to think about problems and arrive at conclusions independently, drawing on principles of mathematics they'd learned before in order to 'discover' solutions for themselves—an approach he called *discovery learning*."[9]

With the advent of handheld calculators, personal computer, and smart devices there is no longer a need for the general population to understand mathematics and even science. We take these devices for granted and do not question their output, despite our errors framing the problem ore 'fat fingering' the input.

For those developing AI programs or prompting the input, the construct of new math may enjoy a Renaissance. Most readers are not mathematicians or scientists by training but will use AI to address very challenging daily problem we could only dream of solving even a short time ago. This makes the problem definition all that more important.

It is critical that the problem be properly defined both contextually as well as mathematically. After all, algorithms are the reduction of problem statement to math that software can properly calculate an accurate, reliable, and valid output.

Typically, questions are asked using conventional and often accepted methodologies. **Perhaps, we need to change the question or at least the frame of the ask**. In the science fiction classic Star Trek, Captain Kirk is asked how he defeated the unwinnable question of the Kobayashi Maru simulation Starfleet cadets are asked "I changed the conditions of the test. I got a commendation for original thinking. I don't like to lose."[10]

This process is arguably the most important part of the entire process. For complicated, important issues a multi-faceted team should be formed with appropriate content subject matter experts as well as those knowledgeable in programming.

For this writer, this is *déjà vu all over again*. Organizational IT initiatives have needed this approach since the inception of modern computing.

However, a major red flag—most organizational initiatives fail and sometime in a major way. Major change management has long been an issue. In 2025, a Harvard article described seven reasons why change management strategies failed. The piece discusses change management approach from the 1990s as the solution to today's problems.[11]

1. **Incomplete or Poorly Defined Strategy**—What is the problem change will solve and what is the detailed and supported plan to accomplish change?
2. **Strategy that is Too Rigid and Inflexible**—The military dictum, "No Plan Survives First Contact With the Enemy" (attributed to several sources) is applicable to organizational change as well particularly with such a dynamic change as AI is causing, even at the societal level. Management needs to act accordingly.
3. **Lack of Effective Communications**—Communications with constituents need to be constant, visible, and coming from the top of the organization.
4. **Failing to Identify and Address Resistance**—Some will fear and resist change. Trust may weaken or even disappear. Management must continue to answer the *What's In It For ME* question that every organizational stakeholder asks.[12]
5. **Disconnect between Strategy and Culture**—This an often-overlooked concern when software and technology providers, consultants and management

are 'selling' transformation. Is the culture ready and what can be done to get it ready? We have addressed this issue for decades and an in-depth discussion is in our 2025 book, *Navigating the Data Minefields: Management's Guide to Better Decision-Making*, Chapter 9.[13]

6. **Setting Unrealistic Expectations**—There is a lot of hype about AI. It is the latest IT 'shiny object.' There are some maturation issues associated with this new technology, many of which are addressed in this book. Management must be careful setting expectations of all stakeholders, including customer and supply change partners.

7. **Not Creating and Celebrating Short Term Wins**—Lessons from project management include near-term or quick wins to document success for team members, employees, management, and all other observers. People look for immediate returns and showing quick success with Phase I of implementations is critical, not just highly recommended.

In 2022, we penned a blog, *Why Corporate Initiatives Fail* and addressed this issue in some detail. We quote a 2019 Forbes Survey that found, "When participants in our survey were asked to create a list of reasons for (change programs) failure, 'insufficient budget' was cited by 23% and 'insufficient time' by only 17%. Instead, participants ranked poor communication (62%), insufficient leadership and support (54%), organizational politics (50%), lack of understanding of the purpose of the change (50%), lack of user buy-in (42%) and lack of collaboration (40%) as the most critical issues." We referenced a 2011 article in the *Journal of Change Management* that argued that organizational change management failure rate is around 70%.[14]

These numbers remain abysmal, and the weakness defined in 2025 remain basically the same as they have been at least since the 1990s. Not much has change in this person's career, initiatives continue to be disastrous and even career ending.

Think of the issues employees are concerned with. Will my job go away? Resistance, even outright sabotage, is a real risk and many of the reasons for failure listed directly relate to this one issue!

With trillions of dollars of AI 'spend' publicly committed by major players, the risk to shareholder value can be enormous if AI projects enjoy this same level of success.

3.3.2.2 Prompting Techniques

In order to input The Problem Statement into a Large Language Model (LLM), we must generate prompts. These can be defined as questions, statement or command, sample code, or other texts. Some support non-text prompts such as images or audio files, and they are the foundation of the AI model response.

However, AI responses to prompts can be inconsistent, incorrect, or even silly. To develop a more effective response, AI Prompt Frameworks are under development. The intent is to make this job easier. Yet, these efforts generate their own issues.

One of the problems AI users have is the plethora of prompts and associated frameworks. A whole book could be written on this subject and by its publication date it most likely would be out of date. Therefore, it seems appropriate to review this subject from the perspectives of frameworks and how to assess applicability to the particular problem at hand. We then review a few of the more popular frameworks to better help readers understand how and when any given one should be selected.

Prompt frameworks are methodical approaches that help the human express his or her intent in context with an output format that is very useful to us humans. Frameworks should be used by individual as well as the largest global organization enabling:

- **Greater Efficiency**—The development of a *library of prompt templates* is high value saving time and money on the large number of competitive task most perform on a daily, even hourly basis.
- **More Accurate**—Well-structured prompts "produce more relevant and factually correct responses."
- **A High Level of Consistency**—Greater uninformative across different teams and projects, including third-party contributions such as contractors, and 1,099 individual contributors.
- **Scalability**—Enable deployment at the enterprise level.

Prompts that meet individual and organizational needs are not a one-size-fits-all, best practice implementations. Their real value comes from solving YOUR PROBLEM! Things to consider when tailoring one of the many prompts templates available include:

- **Define Your Goal**—Be clear and specific. What are we trying to accomplish with a prompt?
- **Analyze Past Prompts**—This is a continuous learning process and determine what works, what does not work, and where improvement can be made.
- **Identify Key Elements**—Why did the work? In what context and how clear are they?
- **Document and Refine**—An actional, reusable framework is the end goal of this effort. Also remember the roles and responsibilities of users, i.e., Business, Developer, Education.

Finally, avoid being too vague, excessive instruction, ignoring context and skipping examples. The so-called AI Prompt Framework 'paperwork' is critical, and the job is not done until this is finalized, reviewed, and approved.[15]

There are a lot of prompt frameworks, and the number seems to be growing exponentially. Table 3.1 is a list taken from several sources listing a selection of representative prompt frameworks in 2025. In some cases the definition is expanded by others and is so cited. This is by no measure a comprehensive list and represents the perspective of this author and other sources as noted.

A few of these prompt frameworks fall into the domain of Prompt Engineering. In the next section we describe this further and the differences between a prompt and prompt engineering.

This area of AI is growing rapidly, and users are encouraged to do their homework. There may be a new prompt framework that *best fits* your problem-solving efforts.

TABLE 3.1
Representative Prompt Frameworks

AI Workflow Studio Top 17 2025 Frameworks[16]

RTF Framework: The Simplest Entry-Level Framework

APE Framework: Essential All-in-One Framework for Beginners

CHAT Framework: Professional's Choice

ROSES Framework: Essential for Project Management

LangGPT Framework: The Future of AI Prompt Programming

Google's Prompt Engineering Best Practices

SCOPE Framework: Strategic Planning Tool

TRACE Framework: Task Breakdown Expert

SPAR Framework: Problem-Solving Expert

CARE Framework: User Experience Designer

COAST Framework: Strategic Execution Expert

RACE Framework: Rapid Action Framework

RISE Framework: Execution Enhancement Tool

SAGE Framework: Wise Decision Assistant

TAG Framework: Concise and Efficient Framework

CRISPE Framework: Creative Generation Framework

SMART Framework: Goal Achievement Framework

8 Powerful Prompting Styles[17]

ReAct (Reason + Act): "Prompting is a technique used to improve AI models in solving problems. It combines two important processes i.e., reasoning (thinking through the problem) and acting (taking actions based on that thinking). This technique helps AI to take more intelligent and adaptive decisions in dynamic environments like robotics, customer service, and decision-making systems."[18]

Chain-of-Thought (CoT): "A prompt engineering technique that enhances the output of large language models (LLMs), particularly for complex tasks involving multistep reasoning. It facilitates problem-solving by guiding the model through a step-by-step reasoning process by using a coherent series of logical steps."[19]

Tree of-Thought (ToT): "A framework that generalizes over chain-of-thought prompting and encourages exploration over thoughts that serve as intermediate steps for general problem solving with language models."[20]

Self-Ask With Search: "The creators of Self-Ask explored how much of a Large Language Model (LLM) correct answers are due to reasoning between facts rather than just memorization. Their intuition was that prompting a model to ask follow-up questions to break down the initial query would enhance its performance on reasoning tasks."[21]

Role-Play: "Role-based prompting is a prompt engineering technique where you explicitly instruct an AI to assume a specific role, persona or character when generating responses."[22]

Few-Shot: "Few-shot prompting refers to the process of providing an AI model with a few examples of a task to guide its performance. This method is particularly useful in scenarios where extensive training data is unavailable."[23]

Reflexion (Self-Critique): "A technique encourages the AI to analyze, revise, and improve its answers through structured self-evaluation."[24]

Maieutic (Socratic Prompting): "questioning technique based on the Socratic method that helps draw out knowledge and understanding through structured dialogue rather than direct instruction. It uses carefully crafted questions to guide learners toward discovering answers themselves, making it a powerful tool for teaching, coaching, and AI interactions."[25]

Treat the selection of a framework as you would any technology assessment and adoption decision (Section 3.3.4 'Technology Readiness Assessment' of Chapter 3). It is as important as any decision made in your AI space and implementation process.

3.3.2.3   Prompt Engineering

Emerging as one of the critical aspects of AI, "Prompt engineering is the process of creating effective prompts that enable AI models to generate responses based on given inputs." In the case of text-based tasks, it is a function of crafting the right instruction set to generate the desired results. The process involves:

- **Crafting the Prompt**—This can be a statement, question, example, etc. specifying what the LLM to do, along with guidance for the desired response.
- **Understanding the LLM**—Models and best prompts will differ. A deep understanding by your SMEs is necessary.
- **Refining the Prompt**—This can be a trial-and-error process. This may be an ongoing process as your business and technology environments change.[26]

This is another major AI dynamic field. The following Prompt Engineering Best Practices list is abbreviated due to space constraints. Additional details are available from the source citation.[27] These items are pretty straightforward; however, as with all prompts and software, the *devil is in the details*. Care must be exercised when developing actionable prompts from this list (Table 3.2).

Prompts are getting smarter. For example, *Meta Prompting* is an advanced technique that uses prompts to generate, refine, and analyze other prompts rather than as a direct response a user's request. This can enhance the performance of LLMs by developing more dynamic, flexible, and effective prompts.[28]

There is also a rich body of knowledge available at **Prompt Engineering Roadmap** (roadmap.sh). Those wanting to do a deep dive into roadmaps, prompts,

---

TABLE 3.2

Best Practices for Prompt Engineering

Be Clear and Specific	Specify Response Format	Provide Context
Structure Step-by-Step Instructions	Set Output Constraints	Experiment and Iterate
Use Clear Action Verbs	Ask for Multiple Perspectives or Solutions	Refine with Clarify Questions
Test Different Wording for Better Results	Use Conditional Prompts for Focused Answers	Request for Examples or Case Studies
Be Transparent About Your Expectations	Use Time Frames or Historical Context	Maintain a Balance Between Open-Ended and Closed-Ended Questions
Clarify the Target Audience	Use Creative or Scenario-Based Prompts for Idea Generation	Incorporate Metrics and Data for Analytical Tasks
Utilize Tone and Voice for Personalization	Request Sources or Citations	

and other development tools may want to look at this material. Potential Applications include AI Development, Research, Business, and Personal Development.

Finally, for organizations that routinely use similar sets of prompts, it may make sense to develop a prompt library based on standardized organizational tasks workflows. This represents new Intellectual Property, Organizational Knowledge, and Know How with value derived at least along the lines of patents and other current IP. This is probably the responsibility of the CAIO and their organization in conjunction with legal and other appropriate departments.

3.3.3 MODEL CONSTRAINTS AND LIMITATIONS

In addition to the technology assessment and selection process, of which part is the identification of the right AI solution for the problem to be solved, AI models and in fact all business and technology models have limits. In this section, we discuss some of the boundaries that are required to assure that AI calculations focus on the problem identified and are not impacted by errors and extraneous questions.

This is true for even the largest models and moreover, often the last mile of problem solving takes the most compute resources and time. Readers may be familiar with the Power Curve, or 'the rate of doing work.' In other words, finding the optimal point of operation rather than asymptotically approaching 100% as the power curve flattens. The last 5% Rule where it may not be cost effective to asymptotically approach 100%.

However, depending on the impact of a solution, i.e., medical robotic surgery, one may want to be closer to 100% than a marketing campaign.

- While we are using a significant number of independent and dependent variables, we are not trying to solve world hunger. Therefore, we need to put constraints or limits on (bound) the model. This is an age-old problem we first addressed in our 2015 blog, Bounding the Boundless where we argued, "One of the difficulties of systemic risk management is to put a boundary or constraints around the problem. This is a major aspect of framing the initial challenge one is trying to solve."[29]
- From a scientific perspective,

    Across all science, modelling is our most powerful tool, as *models let us focus on the few detail that matter most, leaving many others aside*. Models also help reveal the typically far-from-intuitive consequences when multiple causal factors act in combination.

- Additionally,

    Getting AI/ML/DL systems to work has been one of the biggest leaps in technology in recent years, but understanding how to control and optimize them as they adapt isn't nearly as far along. These systems are generally opaque if a problem develops in the field. There is little or no visibility into how algorithms are utilized, or how weights that determine their behavior will change with a particular use case or interactions with other technology.

- Moreover,

  > The European Union this week (2021) issued guidelines for AI—specifically
  > including ML and automated decision-making systems—limiting the abil-
  > ity of these systems to act autonomously, requiring 'secure and reliable sys-
  > tems software,' and requiring mechanisms for ensuring responsibility and
  > accountability for AI systems and their outcomes.

3.3.4   TECHNOLOGY READINESS ASSESSMENT

This is always a major decision process and with technology as transformational and
impactful as AI, my belief is to look to other Mission Critical best practice technology assessment models, framework and set of processes.

Many argue that with AI, we must make a leap of faith. I agree organizations
should be taking leadership roles; however, from the tag line on The Rapid Response
Institute Home page, *"Technology Romance must be met with Fiscal Realities."* We
have also posed simple questions to help decision-makers, these include:

- What is the quality of your data and how is it used in decision-making
  processes?
- How much do you and your organization know about the challenges and
  trends in Big Data and associated applications, such as AI?
- How ready is your organization to fund and implement new advanced soft-
  ware/cloud applications?
- Does your organization know how to realize the full and measurable economic value from these technologies?
- What is your organization's appetite for the risks involved from implementing new technologies?
- How will these technologies be used in your organization?

For AI implementation, once this initial frame is on the table, I prefer something
along the line of NASA's nine Technology Readiness Levels (TRLs).[30]

TRL 9—Successful operations, 'Flight Proven'
TRL 8—Tested and demonstrable, 'Flight Qualified'
TRL 7—Prototype proven in actual operations
TRL 6—Demonstration in a relevant environment (Pilot)
TRL 5—Validated in a relevant environment
TRL 4—Partial validation in a laboratory environment
TRL 3—Proof of Concept (POC)
TRL 2—Application formulated
TRL 1—Basic understanding

This is a good framework and with all guidelines, it can be adjusted to meet your
particular situation. Rigor is the critical issues, and this model takes into consideration, the Proof or Concept, Pilot, and Rollout process for software, other technologies, and AI.

The actual software selection process starts with Problem Identification as well as other evaluation aspects presented throughout this book.

3.3.5   SANDBOX

The process of taking a new software version or new release 'live,' or production always has at least one little pucker factor moment. This is the time when what can go wrong, will. To mitigate the technical and market risks associated with this process, organizations have used the concept of a development site for some time. Even this writer's WordPress website has a development site used for software construction processes. NASA and the aircraft industry have used simulators for decades to test software and make sure new releases perform as expected prior to a general release. Simulators are often used to troubleshoot or reproduce real processes as part of investigations and accident reviews.

3.3.5.1   Aerospace Software Issues

According to NASA,

> Of the historical incidents analyzed, 87% were from software acting unexpectedly rather than simply stopping. Rebooting was found to be ineffective to clear erroneous behavior, and only partially effective for silent software. Errors were traced back to the software logic itself in 62% of cases, 13% within configurable data, and 25% introduced through input. Thirty percent (30%) of unexpected software behavior was caused by the absence of software and 20% was due to 'unknown-unknowns.' These findings indicate that to achieve fault tolerance in safety-critical systems, backup strategies must be employed to detect and respond to erroneous software behavior beyond only fail-silent cases, and robust off-nominal testing should be performed to uncover unanticipated situations.[31]

Most believe and the industry track record supports that air travel is very safe. Otherwise, no one would get into a cramped aluminum tube and hurl oneself across the sky at almost the speed of sound, approximately 7 miles above the earth's surface. The International Air Transport Association (IATA) forecast that 5.22 billion passengers will fly on approximately 40 million flights in 2025.[32]

3.3.5.2   AI Testing

AI sandboxes are isolated environments where businesses can develop, test, and deploy AI models without impacting their production systems. This can be a valuable tool for businesses that are looking to adopt AI, as it allows them to experiment with new ideas and identify potential risks before deploying them in a live environment.[33]

3.3.5.3   Article 57: EU AI Regulatory Sandboxes

For those organizations with EU operations,

> Member States shall ensure that their competent authorities establish at least one AI regulatory sandbox at national level, which shall be operational by 2 August 2026. That sandbox may also be established jointly with the competent authorities of other Member States. The Commission may provide technical support, advice and tools for

the establishment and operation of AI regulatory sandboxes. The obligation under the first subparagraph may also be fulfilled by participating in an existing sandbox in so far as that participation provides an equivalent level of national coverage for the participating Member States.[34]

AI regulatory sandbox potential benefits include:

- Reduction in the time and cost of commercializing AI products and services
- Improved regulation AI learning
- Improving the AI development standards, providing synergies between standards and regulatory development
- Enabling better market participation from small- and medium-sized enterprises (SMEs)

Regulatory sandboxes are under consideration in the United States, Norway, Switzerland, and Spain, among others. This regulatory construct appears to have momentum and will likely become part of the regulatory requirements for those selling as well as using AI goods and services.[35]

Chapter VI: Measures in Support of Innovation, Article 57 is part of a larger framework regulating AI and others regulatory bodies may pattern their legal frameworks accordingly. Another regulation member states, and their organizations must adhere to with more coming, most likely.

3.3.6 GUARDRAILS

IBM defines AI guardrails as, "the safeguards that keep artificial intelligence (AI) systems operating safely, responsibly and within defined boundaries." As the name suggests, this help the organization stay on an acceptable AI road without veering off the path into an accident. They encompass policies and regulations, technology controls, and monitoring mechanisms at the governance and enterprise level.

IBM's 2025 report, Cost of a Data Breach Report, stated that the average cost of a breach in the United States was a record USD 10.22 million. They also state that, "Meanwhile, almost every AI-related breach (97%) occurred in an environment without access controls, highlighting how the absence of safeguards can leave AI deployments exposed."[36]

The organization needs to look at this issue in a serious manner. It is not the responsibility of the CAIO or others, it is the responsibility of the CEO and cannot be delegated. Also, don't forget to involve Legal in this process.

NOTES

1. "BrainyQuote." https://www.brainyquote.com/quotes/albert_bandura_788120. Accessed October 7, 2025.
2. "They Blinded Me with Science." The Rapid Response Institute. https://therrinstitute.com/they-blinded-me-with-science/. Accessed October 7, 2025.
3. "Bombe." Britannica. https://www.britannica.com/topic/Bombe. Accessed October 8, 2025.

4. "Operations Management System." The Rapid Response Institute. https://therrinstitute. com/operations-management-system/. Accessed October 8, 2025.
5. "Why firms are merging HR and IT departments." BBC. https://www.bbc.com/news/ articles/cy0w8gvq84xo?at_bbc_team=editorial&at_medium=social&at_link_type= web_link&at_link_origin=BBC_News&at_campaign_type=owned&at_link_id= E7546A48-7455-11F0-B17C-95E1837CC4EB&at_format=link&at_ptr_name=linked_ in_page&at_campaign=Social_Flow. Accessed October 8, 2025.
6. "Stay ahead, lead the future of AI in project management." PMI. https://www.pmi.org/ learning/ai-in-project-management#:~:text=This%20guide%20helps%20PMOs%2C %20TMOs%2C%20and%20project%20managers,Infinity%E2%84%A2%2C%20is %20designed%20to%20help%20you%20achieve%20more. Accessed October 8, 2025.
7. "Top 8 AI Frameworks: Benefits & How to Choose the Right One." lakeFS. https:// lakefs.io/blog/ai-frameworks/. Accessed October 8, 2025.
8. "The Impact of Generative AI on Critical Thinking: Self-Reported Reductions in Cognitive Effort and Confidence Effects From a Survey of Knowledge Workers." Microsoft. https://www.microsoft.com/en-us/research/wp-content/uploads/2025/01/ lee_2025_ai_critical_thinking_survey.pdf. Accessed June 1, 2025.
9. "What Happened to 'New Math'?" Medium. https://medium.com/age-of-awareness/ what-happened-to-new-math-eeb8522fc695. Accessed June 5, 2025.
10. "Star Trek: The Wrath of Khan." Quotes. https://www.quotes.net/mquote/90879. Accessed March 3, 2025.
11. "7 Reasons Why Change Management Strategies Fail and How to Avoid Them." Havard. https://professional.dce.harvard.edu/blog/7-reasons-why-change-management-strategies- fail-and-how-to-avoid-them/#What-is-Change-Management-and-Why-Do-We-Need- It. Accessed June 5, 2025.
12. "Tomorrow's Culture—Today!" The Rapid Response Institute. https://therrinstitute.com/ tomorrows-culture-today/. Accessed June 5, 2025.
13. "Navigating the Data Minefields: Management's Guide to Better Decision-Making." Routledge. https://www.routledge.com/Navigating-the-Data-Minefields-Managements- Guide-to-Better-Decision-Making/Shemwell/p/book/9781032677934#top. Accessed June 5, 2025.
14. "Why Corporate Initiatives Fail." The Rapid Response Institute. https://therrinstitute. com/why-corporate-initiatives-fail/. Accessed June 5, 2025.
15. "Mastering Prompt Frameworks 2025: The Complete Guide to Smarter Conversations with AI." Marco Brivio. https://marcobrivio.com/prompt-frameworks-2025/. Accessed August 6, 2025.
16. "2025 Complete Guide to Prompt Frameworks: 18 Practical Frameworks to Boost Your AI Conversation Efficiency by 10x." AI Workflow Studio. https://aiworkflowstudio. com/prompt-framework-guide-2025/. Accessed February 4, 2025.
17. "8 Powerful Prompting Styles." Charles Hills. https://media.licdn.com/dms/document/ media/v2/D561FAQE5a3_ct0P6ag/feedshare-document-pdf-analyzed/B56Zgxz 9RnHQAc-/0/1753182389575?e=1755129600&v=beta&t=dCJXVw7Yf42eTAwB4z EX5o_UcEdo7jk_5f-QiL6pQ4. Accessed August 5, 2025.
18. "ReAct (Reasoning + Acting) Prompting." geeksforgeeks. https://www.geeksforgeeks. org/artificial-intelligence/react-reasoning-acting-prompting/. Accessed August 5, 2025.
19. "What is chain of thought (CoT) prompting?" IBM. https://www.ibm.com/think/topics/ chain-of-thoughts. Accessed August 5, 2025.
20. "Tree of Thoughts (ToT)." Prompt Engineering Guide. https://www.promptingguide.ai/ techniques/tot. Accessed August 5, 2025.
21. "Self-Ask Prompting." Learn Prompting. https://learnprompting.org/docs/advanced/ few_shot/self_ask. Accessed August 5, 2025.
22. "Role-based prompting." geeksforgeeks. https://www.geeksforgeeks.org/artificial- intelligence/role-based-prompting/. Accessed August 5, 2025.
23. "What is few shot prompting?" IBM. https://www.ibm.com/think/topics/few-shot- prompting. Accessed August 5, 2025.

24. "Reflection Prompting: Improve Responses with AI Self-Reflection." fvivas. https://fvivas.com/en/reflection-prompting-technique/. Accessed August 5, 2025.
25. "Master Maieutic Prompting for Effective Learning." Relevance AI. https://relevanceai.com/prompt-engineering/master-maieutic-prompting-for-effective-learning. Accessed August 5, 2025.
26. "What is Prompt Engineering - Meaning, Working, Techniques." geeksforgeeks. https://www.geeksforgeeks.org/blogs/what-is-prompt-engineering-the-ai-revolution/. Accessed August 6, 2025.
27. "Prompt Engineering Best Practices for AI Models." geeksforgeeks. https://www.geeksforgeeks.org/blogs/prompt-engineering-best-practices/. Accessed August 5, 2025.
28. "Meta Prompting." geeksforgeeks. https://www.geeksforgeeks.org/artificial-intelligence/meta-prompting/. Accessed August 6, 2025.
29. "Bounding the Boundless." Governing Energy. https://therrinstitute.com/wp-content/uploads/2018/10/Bounding-the-Boundless-May-18-2015.pdf. Accessed October 8, 2025.
30. "Technology Readiness Levels." NASA. https://www.nasa.gov/directorates/somd/space-communications-navigation-program/technology-readiness-levels/. Accessed October 8, 2025.
31. "Software Error Incident Categorizations in Aerospace." NASA. https://r.search.yahoo.com/_ylt=AwrFFoKSfhVoyns5.wYPxQt.;_ylu=Y29sbwNiZjEEcG9zAzEEdnRpZAMEc2VjA3Ny/RV=2/RE=1746267923/RO=10/RU=https%3a%2f%2fntrs.nasa.gov%2fapi%2fcitations%2f20230012154%2fdownloads%2f8-17-23%252020230012154.pdf/RK=2/RS=bhrPO0PpbAjI74BcnpcmG.ZcZTQ. Accessed May 2, 2025.
32. "Number of air travelers to cross 5 billion for first time in 2025; average fares to drop: IATA." The Straits Times. https://www.straitstimes.com/world/number-of-air-travellers-to-cross-5-billion-for-first-time-in-2025-average-fares-to-drop-iata. Accessed May 2, 2025.
33. "How to Set Up AI Sandboxes to Maximize Adoption Without Compromising Ethics and Values." Medium. https://medium.com/@emilholmegaard/how-to-set-up-ai-sandboxes-to-maximize-adoption-without-compromising-ethics-and-values-637c70626130. Accessed October 8, 2025.
34. "Article 57: AI Regulatory Sandboxes." EU Artificial Intelligence Act. https://artificialintelligenceact.eu/article/57/. Accessed May 5, 2025.
35. "Towards an AI regulatory sandbox: Emerging research and pilots." AI Standards Hub. https://aistandardshub.org/toward-ai-sandbox. Accessed May 5, 2025.
36. "What are AI guardrails?" IBM. https://www.ibm.com/think/topics/ai-guardrails. Accessed October 8, 2025.

4 Risk Mitigation and Enterprise Alignment

The great tragedy of science—the slaying of a beautiful hypothesis by an ugly fact.

—**Thomas Huxley**[1]

There can be a lot of trepidation when implementing new technologies at the enterprise level especially when the probable transformation will be huge. However, individuals need not be frozen in fear about artificial intelligence (AI) implementation. Similar words, hype, and sales efforts are familiar. Reengineering of the 1990s, Y2K, ERP, CRM, SCM, and even the Internet are recent examples. The consulting and research firm, Gartner even has a name for the technology adoption lifecycle—the Hype Curve.[2]

We have been using stochastic decision models for over 30 years. Our belief that linear point A to point B solutions do not reflect the real world. We developed an approach to probability coined as probability density.

Risk management is the core competency of any firm regardless of sectors served. By definition, risk implies a level of uncertainty that must be mitigated. In other words, organizations must reduce the PROBABILITY DENSITY of a continuous variable.

Calculus is a convenient tool as the area under a curve is function of either value or risk.[3] Specifically, a histogram is a graphical way to depict the probability distribution associated with a continuous random variable. Increased knowledge is one way to decrease histogram area or range of uncertainty. The concepts of uncertainty and the management of this risk will be developed later in this document.

Lower costs do not necessarily provide competitive advantage. A low-cost structure is simply a facilitator that must be capitalized on. What an organization does operationally will ultimately determine available free cash flow that can be translated into greater returns to shareholders and a higher stock price.

There is an optimal envelope (OE) within which an acceptable range of returns adjusted for risk is satisfactory. The concept of Pareto optimal and the alignment of metrics were previously developed. For now, assume that this OE can be determined using financial and economic models. Building from the concept of risk management, the OE is understood to be a range within a probability density where risk is mitigated, acceptable, and the likelihood of an acceptable return is satisfactory.

Risk also includes product liability or harm based on AI use, misuse, or errors and omissions. A full enterprise AI risk profile needs to include all known risks with provisions for unknown or new risks. It will be a work in progress throughout the AI life cycle and will set the stage for derivatives and post AI new technologies.

DOI: 10.1201/9781003646914-4

4.1 AVAILABLE DATA AND INFORMATION

A review of the available materials on Artificial Intelligence Technology Suite, including Machine Learning and even Virtual Reality, Augmented Reality and Smart, etc., yields a vast amount of information. Often conflicting, there are at least five types of written (today most are available electronically) materials including text, data, and mathematics:

- **Peer Reviewed**—Authored by credentialed individuals and generally academic by nature. A wealth of documented materials that are usually cited.
- **Books Based on Documented Research**—Books such as this edition in all media delivery formats.
- **Books by on Relevant Subjects**—Often published by university publishers with established credibility but not well researched. They may even be the first time a subject is broached, in all media delivery formats as well.
- **Industry and Other Publishers**—Often in a magazine article format. Reports by established industry consulting firms or think tanks.
- **Blogs and Other Opinion Pieces**—Common online and often articles posted on social media sites.

This list is not intended to be all inclusive and other sources of information and data are available as well. The intent is to frame data and information systems that may lead to others forming an opinion and even knowledge.

AI and its derivatives are all the rage. Readers can find significant written material in all of the above listed media. There is so much and the intensity so great that it is hard to decern and categorize the value of individual publication.

4.1.1 D-K QUADRANTS

The result of a 1999 research paper by social psychologists David Dunning and Justin Kruger, "The Dunning-Kruger effect is a psychological tendency in certain people who perform tasks poorly yet overestimate their knowledge and abilities compared to others. It is about being ignorant about their own ignorance." This bias can lead individuals to incorrectly believe they are more competent and smarter than others than they are in reality.[4]

From the standpoint of new technologies, it appears that *irrational exuberance* about the new cool stuff is the first step. Early advocates gain a knowledge foothold and become the equivalent of the local expert. Many of these individuals reach erroneous conclusions and often their incompetence robs them of the metacognitive ability to realize it.

We see this effect now, with a plethora of individuals claiming significant expertise and offering large numbers of commercial solutions in this new arena. Eventually, those interested in the subject become more aware and the impact of these pseudo-experts diminish.

This effect is often treated as the basis of the technology maturation curve and appears to mirror the Gartner Hype Cycle.[5] Regardless, it is useful to view technology maturation through a similar lens.

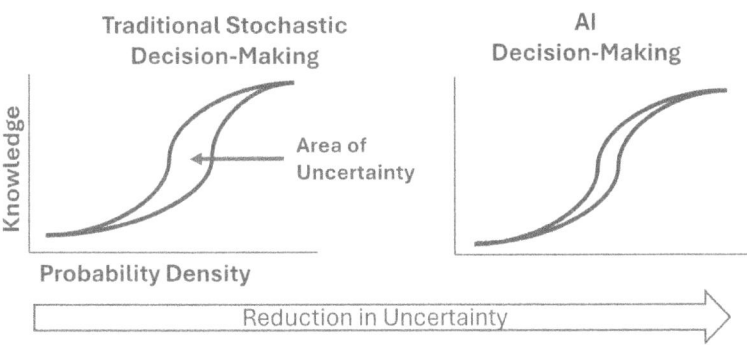

FIGURE 4.1 AI-driven risk mitigation.

Empirical evidence suggests that a technology body of knowledge will grow along two axis, a growing real confidence in the understanding of technology and a similar enhancement of overall expertise. By extending the Dunning-Kruger effect, we visualize four D-K Quadrants as shown in Figure 4.1. Quadrant models enable the simultaneous position of variables as opposed to a somewhat subjective positioning on a life cycle curve.

On the vertical axis, Confidence, the scale is from Faux or Pseudo to a High level. This reflects the process that unfolds with the new and unknown, including technology as it gains general knowledge and acceptance. On the other hand, the horizontal axis Expertise is positioned between Low and High levels.

We define the four quadrants as follows:

- **Irrational Exuberance**—The new technology reaches a critical mass whereby it is either commercial or a minimum viable product (MVP) and many see the value in being seen as an expert. Social media feeds this model and while there are serious economic players, there are many *want-to-be* and even charlatans. Many true advocates and/or evangelists, knowledgeable or otherwise emerge at this stage.
- **Arrogance**—This is a selfish, boastful trait by those that use this narcissistic quality to shield their insecurities. The exaggerated sense of importance leads these individuals to believe they know it all, have no interest the opinion of others, and will take ridiculous positions to make the point. Arrogant prognosticators taint the point/product/argument they are making with their bias. The quality of their assessment is suspect.
- **Self-Awareness**—"Involves recognizing one's own emotions, strengths, and weaknesses, which enhances personal and professional growth. Improving self-awareness can be achieved through mindfulness, feedback from others, and reflective practices like journaling. High self-awareness leads to better

decision-making, improved relationships, and increased emotional intelligence."[6] While their view of AI products may lack certain expertise, they are confident enough in their beliefs that they can acknowledge weaknesses.

- **Guru**—This is the quadrant of thought leadership by true experts. Typically, AI is mature and understands by a critical mass of those with significant experience and capability in the field. The trust of these individuals/organizations is rightfully high.

While we mentioned the lower left quadrant as the area many currently inhabit. In this model, it is possible for individuals and entities to inhabit areas of all four quadrants simultaneously. This is part of the strategic discussion herein, as real value is not attained by those with irrational exuberance. Case studies and AI models we describe are found throughout the spectrum.

The Technical Maturity foundation is a function of the skills of the talent pool available or will be available in the near term as well as the leaderships cadre of individual entities such as the use community. The leadership exhibited by key technology players is critical as well. Without these two components, technological development, growth, and maturation are rudderless leaving these technology suites adrift. We have seen many promising tools that never get real traction, for example, expert systems, real options, and certain operating systems.

Some technologies, for example, AGI and Superintelligence, continue to emerge, but even version 1.0 models may occupy the Guru space. So, not just a maturity continuum along a line, but the desired movement is from lower left to upper right. Unfortunately, some may deviate into Arrogance, while Self-Aware may be a brief rest stop. New firms and AI technologies will likely emerge and most likely follow the same vector (Figure 4.2).

The approach taken with this model is consistent with the Rapid Response Management Matrix model first documented in 2009.[7] In that model, we defined each quadrant as an allegory of athletic skill complete with a definition and set of traits. The intent was to help readers better relate to entities that were typically well known. This has been our intent with the D-K Quadrants model.

As with quantum physics, and this model, AI maturity can occupy more than one space simultaneously. This is a clearer picture of fast-moving technology development with a large number solution suites and derivatives.

4.1.2 BUYER BEWARE

My personal experience researcher for this book includes all sources from all of the above cited as appropriate herein. One of the more revealing finds is that the preponderance of AI (opinions) comes from sources that are seemingly not well researched or even supported.

Sometimes, even copied from each other. Perhaps plagiarized; however, it is not clear who the original author(s) are. Blogs, Videos, LinkedIn, and Substack are the majority of reading I have done from this class of publications.

A multitude of experts are emerging, very similar training, and credentials are offered by a number of organizations, and graphics, templates, flowcharts, charts,

FIGURE 4.2 D-K quadrants.

and solution/software recommendations, etc. are freely reproduced. Some self-proclaimed experts sound good, but several have not passed a more rigorous review of their credentials. One wonders if they are compensated for clicks, likes, etc.?

In several cases, some materials seem made up. One individual appeared to take a documented press release by a large public company and elaborated on the project management process and personnel requirements offering judgment to the effect 'this is amazing.' I could find no other reference that went into that detail and most stated, words to the effect, "details were not released," which is a common practice of publicly traded firms.

There is a lot of excitement about AI, and it will likely continue as have other new technologies. Most authors are not altruistic including this one. Point being, a lot of money is at stake, and this always attracts attention.

4.1.3 AI-Washing

The saga of London-based startup, Builder.ai is another cautionary tale. Apparently, they sold investors and customers, including Microsoft and Amazon they could produce an app entirely using AI. In an extreme case of *AI-Washing*, were businesses/individuals benefit from the new technology excitement and hype without actually using it, the firm reportedly hired 700 Indian engineers to pose and do the work of its AI chatbot, Natasha. These misrepresentations cost millions and ended in bankruptcy with the resulting layoff of approximately 1,000 employees.[8]

A concern is that the AI-washing practice will continue despite Security and Exchange Commission (SEC) fines because in this hyper AI market, the temptation for illicit operators is very high.[9] Investors and buyers need to perform reasonable due diligence before developing a financial relationship with start-up firms in this hypermarket. The same goes with the possible merger or acquisition of these firms.

4.1.4 Model Collapse

Recently, researchers have found that generative AI (GenAI) models trained solely on predecessor model outs produce increasingly inaccurate results. Effectively, AI *inbreeding* which produces weak new models that themselves are prone to errors. Model collapse has serious ramifications, especially on Mission Critical applications. These points are taken from IBM material and include the following[10]:

- **Poor Decision-Making**—Inaccurate AI output can have a devastating impact on everything from customer service chatbots to medical diagnostics.
- **User Disengagement**—Users will abandon poorly performing systems, and this has the potential to put the entire AI initiative at risk generating a potential negative impact on the equity markets and other stakeholder, i.e., customer, regulators, and others.
- **Knowledge Decline**—If loss of confidence ensues, this could lead to a degradation of Knowledge Intellectual Property and disuse of organization's AI solutions, both internal and external.
- **Catastrophic Forgetting**—Akin to human dementia, this deterioration in AI performance will get worse with each model generation. Effectively a loss of organization knowledge and competitive position.
- **Mode Collapse**—This is specific to Generative Adversarial Networks (GANs) models. "Mode collapse occurs when the generator's output lacks variance and this flaw goes undetected by the discriminator, resulting in degraded performance."
- **Model Drift**—"If an AI model's training, based on old training data, doesn't align with incoming data, it can't accurately interpret that data or use that incoming data to reliably make accurate predictions."
- **Performative Prediction**—If AI is trained on old or inaccurate data it can provide human decision-makers with output that may lead to poor decisions and even decision violating ethical standards or law.

This is a major issue, but this risk can be mitigated and should be addressed by any AI governance and risk mitigation processes. Model collapse can be prevented by:

- **Retaining Non-AI Data Sources**—Most data management standard suggests that 'original' high quality primary (raw) data be preserved. Basically, preserving the ability to 'restart' if necessary.
- **Determining Data Provenance**—We addressed this in detail in *Navigating the Data Minefields: Management's Guide to Better Decision-Making*. Additionally, 'The Data Provenance Initiative,' a collective of AI researchers from MIT and other universities that has audited more than 4,000 datasets.
- **Leveraging Data Accumulation**—Continue to train model with both real data as well as the multi-generations of synthetic data.
- **Using Better Synthetic Data**—High quality data, regardless of its source is mandatory for strong AI solutions.
- **Implementing Data Governance Tools**—Strong bond AI governance is mandatory (Section 1.7 Governing AI).

It is important that management understand this issue and AI governance, policies and procedures assure this potential devastating failure be mitigated. AI model collapse can put the company and even lives at risk.

The CAIO should be held directly responsible for this risk. This issue should also on the calendar of quarterly Board of Director Risk Committee.

This is a bet your company and bet your career issue!

4.2 NEAR-TERM RISKS VS LONG-TERM RISKS

The West, and particularly the United States, tends to focus on the near term, aka immediate gratification to problems or material desires. Long term is not part of the corporate quarterly reporting mentality as much as individuals forgoing long-term consequences to current actions and decisions. This *Instant (Immediate) Gratification Bias* often clouds our judgment. This tendency is inherent to our psyche and is also known as the pleasure principle. "The term was first used by Sigmund Freud to describe the role of the 'id,' his proposed component of the unconscious mind that is driven purely by baser instincts."[11]

Most people are vaguely familiar with certain risks posed by AI, with one of the most articulated "will I lose my job" concerns. Risks can run even more personal and catastrophic. One example, the bipartisan TAKE IT DOWN Act was signed into law on May 19, 2025. It is the result of broad advocacy after a 14-year-old female's image was modified into pornography using AI and distributed on social media without her consent. This law provides certain legal remedies for this type of behavior.

There is general agreement that AI poses risks. First, the existential risks that humans cannot control are unforeseen and potentially catastrophic.

A small group is concerned that AI will achieve human-like intelligence that results in the extinction of humanity.[12] This includes Nobel Laureate and AI pioneer,

Geoffrey Hinton, who has expressed concern that there is a 10%–20% risk that AI will eventually become our overload. A one in five chance? To this pundit, that is a significant number.[13]

Other more immediate concerns include the risk of societal biases being embedded into AI systems (Section 1.5.2.2 'Debiasing' of Chapter 1).

Pornography is another thorny AI real issue (Section 1.11.1 'Will AI Replace Me?' of Chapter 1) the AI sector faces. Not just those in the adult entertainment business but the (criminal) capture of real people likenesses for nefarious purposes.

Additionally, some are alarmed by the outsize control AI software companies may have directing our behaviors and especially with our children.[14]

Everything new has a risk profile that will continue to evolve. There are gaps or unknowns in our knowledge base. Some will be filled, and new ones will open. Organizations need a robust risk mitigation process, as with any strategic core competency necessary for success in todays and future markets.

4.2.1 Medical Superintelligence 101

Is medical superintelligence right around the corner? According to Microsoft, the company has developed "MAI Diagnostic Orchestrator (MAI-DxO) that queries several leading AI models—including OpenAI's GPT, Google's Gemini, Anthropic's Claude, Meta's Llama, and xAI's Grok—in a way that loosely mimics several human experts working together." Essentially an orchestration mechanism, the system used multiple AI agents that "loosely mimic several human experts working together." Microsoft indicates that Microsoft AI Diagnostic Orchestrator (MAI-DxO) achieved an accuracy of 80% versus 20% for doctors. It also reduced the cost of treatment by 20% by using less expensive tests and procedures. This *chain-of-debate* style of multiple agents working together is driving the company and by extension society closer to medical superintelligence.[15]

Each week, the New England Journal of Medicine (NEJM) publishes a case record from the Massachusetts General Hospital in a detailed, narrative format. "These cases are among the most diagnostically complex and intellectually demanding in clinical medicine, often requiring multiple specialists and diagnostic tests to reach a definitive diagnosis." Using 304 case studies from NEJM, the researchers developed a Sequential Diagnosis Benchmark (SDBench), a stepwise iterative solution that mimics doctor's (realistic clinical) diagnoses. The diagnosis can be compared to the 'gold standard outcome' published in the NEJM. Finally, a virtual cost is assigned to each investigation that is modeled using real world expenditures. The quality of the medical diagnoses and cost and use of resources is evaluated.[16]

There are current and growing concerns about medical care availability and cost. These AI-powered diagnostic service can bring low-cost, world-class quality medical care to rural and remote areas throughout the world as well as second opinions on complex challenging cases. Perhaps coupled with remote robotic surgery, this can be a force multiplier for this sector and the global 8+ billion potential patients.

This collaborative model should have value in many other critical and complex applications. Smart systems and neural networks come to mind as do power grid optimization to name a few. We have known for years that small high-performance teams perform better than single subject matter expert.

4.2.1.1 What If AI Is Wrong?

Throughout this book and our previous book about data management, we address pros and cons as well as hypothetical and use cases where AL/ML has not performed satisfactorily. Since this technology is early in its maturity and major complex problems are just getting started, how do we determine if the answer is correct, its accuracy, margin of error, bias, or completely incorrect?

We distinguish from AL hallucinations, *a phenomenon wherein an LLM perceives patterns or objects that are nonexistent or imperceptible to human observers, thus creating outputs that are nonsensical or altogether inaccurate. By wrong we mean inaccuracies (see Appendix I: Glossary of Terms).*

A recent study led by researchers at the Icahn School of Medicine at Mount Sinai, working with colleagues from Israel's Rabin Medical Center and other institutions, has found that even today's most advanced artificial intelligence (AI) models can make surprisingly basic errors when navigating complex medical ethics questions. The results, published online on July 22 in NPJ Digital Medicine, raise important concerns about how much trust should be placed in large language models (LLMs) like ChatGPT when they are used in health care environments.[17]

We will discuss the AI methodology and Problem Statement (Section 3.3.2, 'Problem Statement' of Chapter 3) as well as other tools. It is important that the human maintains a high level of skepticism about AI prompts, training, use, etc.

- Quality and specificity of prompts
- Nature of the text output
- Benchmarking
- Extensive testing before going live or to market
- Consistency of results
- 'Gut' feeling

Individuals who have made a career in certain segments have a 'gut feeling' for what the output should be. You may find these people in your organization or supply chain. However, there is another source—retired professionals.

Many can be retained to review the solution output. I personally know a retired pathologist who has litter reviews thousands of tissue samples. That person's gut can be very valuable.

NASA has developed extensive testing procedures for AI, software, critical systems, and other technologies, including third-party audits.[18] Others may include providers in other critical infrastructure sectors. Benchmarking and adapting best practices may be an effective way to address error concerns.

However, organizations are not 'apples to apples,' so be cautious when using benchmarks and best practices even from your sector. Every organization is different with a different culture and management style.

4.2.2   IS YOUR AGENTIC AI PROJECT AT RISK?

"Over 40% of agentic AI projects will be canceled by the end of 2027, due to esca-lating costs, unclear business value or inadequate risk controls." Gartner goes on to state that most agentic AI are either early stage or even proof of concept (POC) that are driven by hype and often used in inappropriate applications. Organizations often do not have a good handle on the real cost and difficulty deploying AI at production operational scale. They need to be more careful about their assess-ment of the strategic alignment with the business (Section 1.7 'Governing AI' of Chapter 1).[19]

Gartner also states that only about 130 of the thousands of agentic AI providers are real. Agent washing, aka rebranding of existing products without substantial value added, remains a problem. Moreover, the existing technology does not possess the capabilities assigned to it by the 'hype.'

Rethinking workflows using agentic AI needs to the foundation of the AI-enabled business. Value can be attained through cost management, high performance, qual-ity, speed, and enterprise level scale (Section 1.10 'Value Derived' of Chapter 1). Agentic AI can provide greater resources efficiency, automate tasks, and enable new business innovation.

By 2028, Gartner expects at least 15% of the daily decisions will be made autono-mously from zero in 2024. Additionally, from less than 1% in 2024, 33% of enter-prise software will include agentic AI by 2028. Real value from agentic AI is at the enterprise level, not individual tasks, or departments.[20]

Having made a career in the energy sector, where life and death issues as well as the enormous costs of technology implementations cause management to tread lightly in this risk adverse environment. Agentic AI is one of those technol-ogy transformations where careful risk assessment and cautious deployment is warranted.

Bottom line—vendor selection is just as important as technology selection.

4.2.3   EXISTENTIAL, EXISTENTIAL, EXISTENTIAL

In the current environment, advocates routinely scream the term 'existential' over every perceived wrong or offense. So, if everything is existential is anything exis-tential? Like the *Boy Who Cried Wolf*, we are lulled into insensitivity. To date, these existential threats (at least in the near term) have not materialized. This further jades our consciousness about a subject of such great concern.

According to the late United States Secretary of Defense, Donald Rumsfeld,

> There are known knowns. These are things we know that we know. There are known unknowns. That is to say, there are things that we know we don't know. But there are also unknown unknowns. There are things we don't know we don't know.

In our fast-moving technological climate, unknown unknowns will manifest them-selves sooner rather than later, although new unknowns and unintended consequences will emerge. How individuals and organizations position themselves accordingly will help separate winners and losers in the great 'AI lottery.'[21]

Societal transformation enabled by technology is not new and as always, long-term ramifications are not well understood or even articulated by those at the advent of technology advances. The crystal ball into the future of the Internet was cloudy in 1995 at the beginning of the dotcom bubble (Chapter 1 'Introduction'). Appropriate governance is one way to 'future proof' uncertain/unknown risks in the current period.

4.2.4 WE HAVE BEEN HERE BEFORE

In the late 1980s song, The Living Years, the first verse goes, "Every generation blames the one before, And all of their frustrations, come beating on your door."[22] It is meant as regret about the relationship with a deceased parent and now the inability to understand each other. Another interpretation is that now the adult child knows what the parent was saying to his/her child.

AI maturity is similar. After all, it is mimicking human behavior. We and later AI learn from those before us, hence the term 'standing on someone's shoulders.'[23]

Lots of new things have threatened humanity. Changes were made and we, the humans took the leadership and for the most part, harnessed technology for our benefit. There is no reason to believe AI is fundamentally different. The collective we just need to be smart about this.

4.2.4.1 Scaling

The concept of moving from POC through Pilot then to the enterprise level is a challenge made more difficult in a fast-moving environment like AI. According to IBM, these steps will help the organization scale effectively.[24] They are not much different from large projects of any technology, so most will be familiar with them.

- **Start with Data Science**—Work with appropriate Data Science and Machine Learning subject matter experts (SMEs) to develop algorithms and APIs appropriate to your business needs.
- **Locate and Ingest Data Sets**—Putting together the right high quality is mandatory for training AI models.
- **Involve Stakeholder across Department**—It is critical that all stakeholders engage with process. This is not a technical project, it is a business project, and the business segments are the owners.
- **Manage the Data Lifecycle**—This is a major challenge as the data lifecycles are long with many data and technology changes.
- **Optimize and Simplify Machine Learning Operations (MLOps)**—Assure that the MLOps platform is aligned with the skill set of your data science and IT team and supports your IT infrastructure and technology set.
- **Assemble a Cross-Functional AI Team**—This is mandatory for today's organization, especially AI which is driving business behavior as opposed to capturing and report past action. Similar to real time and telemetry key engineering and business, owners must be directly and responsibly involved.
- **Select Projects with High Success Potential**—Initially focus on high visibility, high impact, and quick projects. This shows success and builds buy-in and momentum as people want to be associated with winning.

- **Incorporate Governance and Compliance**—The importance of AI governance as a critical aspect of organizational governance cannot be overstated. AI is the way the company is run, not a bolt on as IT often is. Regulatory compliance and IP management are critical components.
- **Employ the Right Tools**—Equip data scientist, IT, engineers, and others with the tools and the right environment to accomplish their tasks.
- **Monitor AI Models End-to-End**—Track and metric real-time Key Performance Indicators (KPIs) that enable any course correction quickly and helps optimize performance as the project evolves.

This is just good business practice. Management should expect AI to become align with the firm and the firm to align with AI. This has been the case for most if not all new capabilities, i.e., smart devices, manufacturing, and delivery services. Generations Alpha and Beta will see AI with the same lens that Gen Z sees smart devices.

4.2.5 ECONOMIC VALUE

I am surprised that many AI Return on Investment (ROI) models are pretty loose and more of a wish than an actual analysis. This is disappointing since we have known how to do a sophisticated financial and economic analysis for decades. This section will address this process and offer a solution that the Chief Financial Officer (CFO) will love.

In July 2025, I read a Substack piece, the gist of, many measure AI costs and value from the standpoint of the technology and its implementation. The total life cycle of AI capital investments are ignored. However, in my experience, many overlook this type of analysis because it can be difficult and time consuming as well as outside the expertise of many business analysts. This statement also applies to major business and technology professional services who market technology ROI solutions and expertise.

Steve Jobs of Apple is credited with saying,

> Some people say give the customers what they want, but that's not my approach. Our job is to figure out what they're going to want before they do. Henry Ford once said, 'If I'd ask customers what they wanted, they would've told me a faster horse.' People don't know what they want until you show it to them. That's why I never rely on market research. Our task is to read things that are not yet on the page.[25]

This pundit's interpretation of the famous quote is not that market research is of little or no value, but in the emerging technology space, people do not comprehend game changing technology. Therefore, they cannot put an economic value on the solution. AI may currently be trapped in this paradigm.

Once again, organizations are told that they must reinvent themselves, transform the culture, and that their IT legacy capital commitment of not that many years ago is no longer advantageous and is destroying their competitive posture and shareholder value. This story has been heard by the 'C' suite throughout those individual's 30 plus year career. Most value propositions put forth are lame and without merit. They are often *overstated*, not *demonstrable*, or *defendable*, or do not answer the question, 'what's in it for me' if we make this investment?[26]

We have argued for over two decades that the economic value emerging technologies, including tangible and intangible elements can be measured and demonstrated to decision-makers that it meets the test of positive value to the enterprise. This section addresses this well-established approach as it relates to AI's significant capital, bet your company (and career) investments.

4.2.5.1 Thought Leadership ROI

We live in an era of 'influencers,' those who can command a large following of clicks and likes. Yet during the COVID-19 pandemic, physicians and other very well-qualified medical researchers were shunned and often at great personal and career expense. Moreover, organizations are quick to retain outside third-party experts whose advice is taken over those of the firm's internal expertise. We are seeing a plethora of self-proclaimed experts in the AI space these days, hence following this historical trend.

What is *Thought Leadership* and how can an organization capitalize on it?

> Thought leadership stems from a person's own mind that pioneer's intellectual property, spawning concepts that inspire and move others into action. They tend to have a finger on the pulse of enterprise, creating a brand that promises to foster not just communal trust but universal subscription. Innovation is an integral part of thought leaders, raising the eyebrows of not only their own establishment but those of customers. They send industry arch-rivals scrambling to the drawing board to revisit their own blueprint in the quest to keep up with the play.[27]

In an interview with IBM executives by the consulting firm, McKinsey focus on points regarding the ROI or business value of content[28]:

- Thought leadership is a very hot topic and of significant value and the advent of AI organizations will only accelerate that value proposition.
- Over a three-year period beginning during the COVID-19 pandemic, almost 90% of executives surveyed as a result of their understanding of the thought leadership in their area of organizational operations.
- Thought leadership is believed to drive value in five areas: driver of revenue and profit, source of competitive advantage, improves innovation and agility, drives employee satisfaction, and closes knowledge gaps helping organizations compensate for poor data quality and weak analyses.
- Executives use knowledge gained from thought leadership in procurement processes. Expanding the results of this survey, the total yearly procurement spending for the combined US government and Fortune 500 is conservatively estimated to be in excess of $100 billion, and $265 billion for Fortune Global 2000. Significantly, more as the procurement spend base is expanded to include smaller organizations and government entities.
- A successful thought leadership strategy is a function of quality dad and insight, a unique competitive point of view, the ability to reach the intended audience, perception of independent thinking, and trust.
- Finally, leaders must pull the stakeholders toward their vision. The hard sell will usually be met with increased resistance.

The authors estimate that the ROI from thought leadership is on average 156%. This calculation is made using an approach similar to the author's Economic Value Proposition Matrix (EVPM) model (Section 4.2.5.2.1 'Economic Value Proposition Matrix').

AI inference engines thrive on high-quality content, both internal as well as from third-party sources. Therefore, content is a foundational element in the AI era.

4.2.5.2 AI Value: Enable Measurable Business Gains

New technological advances are always met with a great deal of optimism and hype. The latest 'sliced bread' is going to change the world, and proponents often over-state the value. This is especially true when many product/service providers stand to make, in some cases billions of dollars and uninformed technical and executive management fawn over what might be.

This author has long believed that ALL capital culture changing investment must meet the same Capital Expenditure (CAPEX) standards. Their value must be documented and demonstrated. If proponents cannot meet these tests, management is well advised to wait until they can. This section offers use cases that might be used as good practices when determining AI value propositions.

4.2.5.2.1 Economic Value Proposition Matrix

Determining value from investments in information technology programs has and remains difficult to measure, especially for intangible variables such as productivity, safety, employee satisfaction, etc. Over 20 years ago, as part of a Red Team Review for a capital investment in the then emerging Digital Oilfield, we were asked to identify and quantify its value proposition.

We developed a stochastic decision-support model which included five areas of interest with our tag line, "*Translating Your Technology into the Language of Business*." Table 4.1 shows five categories of business value available for IT type investments.

This solution is fully vetted and has been widely used for major investment including security in a geopolitically volatile part of the world. This was documented in a case study with our major defense contractor partner using a predictive model.[29]

The goal is to make value proposition **Believable**, **Demonstrable**, and **Defendable** through the normal organizational capital Authorization for Expenditures (AFE) project scrutiny.

This model and additional case studies is fully documented on our website. Additional information and use cases are provided in *Navigating the Data Minefields: Management's Guide to Better Decision-Making*.

4.2.5.2.2 Translating Time Saved into Business Gains

A 2025 report by the American manufacturing publication Industry Week revealed some interesting concepts as the result of several surveys and other responses regarding returns on AI investments. In the introduction to the article, they comment that while many see AI as the driver of a workplace revolution, there is other value to be captured as well. They caution that productivity gains at the workplace can lead to productivity leakage, when gains at the individual level do not add up to an

TABLE 4.1

EVPM Categories of Value

Category	Definition	Example
Cost Takeout	Completely eliminating a specific activity or process	• Redeploying a resource from a non-value-added activity to a value-added activity
Cost Avoidance	Identifying and correcting an error that was not budgeted for correction but would have caused an expense had it not been corrected	• Correcting an engineering design flaw before the flaw goes into production.
Productivity and Efficiency Gains	Increase in productivity that improves existing resource utilization	• Removal of a bottleneck that is causing capacity restraint • Correcting a process to allow more productive time by shifting from wait time to production time
One-Time Cash Flow Impact	Decreasing and/or eliminating one-time cash flow impact	• Elimination of redundant information/data stores • Monetize Capital
Intangible	Benefits that improve operations of the business and/or are necessary to control, protect, and enhance company assets, but are not quantifiable due to the nature of the area being improved	• Improvement of communications between different operational units/supply chain • Reduced small equipment shrinkage

understandable and measurable business value. This is similar to challenges we have seen with other IT initiatives.

Unless otherwise cited, the key points of the article follow in bullet format.[30]

- According to the Boston Consulting Group (BCG), over 80% of consultants using GenAI believes it enhances problem-solving and generates faster output. However, "Does this translate to real organizational efficiency—or just personal task relief?"
- Gartner's 2025 CEO and Senior Business Executive Survey stated that growth is the top priority for 56% of the organizations and AI is seen as a key enabler. However, while AI saves about 5.7 employee hours per week, only 1.7 hours are used for 'high-value' tasks that improve results. Eight-tenths of an hour (0.8) is spent fixing AI errors. This leaves 3.2 hours per employee unaccounted for.
- Likewise, Microsoft 2025 CEO Study found just while 34% of CEOs expect GenAI to boost productivity, 43% expect improved decision-making. This suggests that organizations are prioritizing the impact from AI rather than simple time saving.
- However, higher productivity benefits can be attained. Gartner reports that 81% realized significant savings at the enterprise level; approximately 27%

higher than less productive organizations in the sector. Moreover, 71% reported stronger innovation resulting in more new products and services.

- Gartner also reports that AI take up is not consistent across organization. For example, 60% of finance organizations still use manual methods. This could be because the nature of financial reporting processes is more conservative or concerns about emerging technologies. In this writer's opinion, legacy memories of the significant problems and cost overruns during Enterprise Resource Planning (ERP) implementations may still haunt senior financial executives.

- Moreover, organizations need to develop a set of metrics that align with their strategic and operational goals. Examples include the following:

  1. **More than Simple Time Saved**—Correlate how personal productivity solutions are used with both the performance of individuals and relevant teams. Measure how AI is leveraged at all levels of the enterprise in support of critical business processes.

  2. **Measure Business Outcomes**—Focus on changes in KPI performance, not simply technology performance.

  3. **Redesign Processes with AI in Mind**—We learned in the 1990s that simply automating existing processes resulted in suboptimal performance. Only when processes were reengineered to take full advantage of the technology suite did high-performance become possible.[31]

  4. **Reskill and Upskill**—BCG discovered that survey participants with moderate coding experience outperform novice on GenAI-augment tasks that did not require coding knowledge. As noted throughout this book, a broader understanding of the context of the problem coupled with requisite experience increased the effectiveness of AI. This suggests that most current engineering, scientific, technical, and even non-technical developers, users, and management will benefit greatly from additional appropriate skills. Moreover, keep skills and knowledge current remains a career necessity.

  5. **Rethink and Define Productivity**—The concept of continuous improvement has been a focal point for successful and high reliability organization. Superior performance is not generated when trivia are among KPIs. Don't sweat the small stuff and continuously examine all aspects, workflows, teaming etc. Strive to attain and maintain Pareto optimality (Section 3.1.1 'Human Input Still Needed' of Chapter 3).

As has always been the case with technology and specifically information technology, enabling business is the only goal. Cool stuff is fine if it aligned with the mission and always remember and reinvent as necessary to meet that goal.

When we first developed EVPM in 2004, we found over 30 opportunities for value when implementing the Digital Oilfield. Some were very high value, some of average value, and some low. Overall, the business unit benefited from most and the sum total from all metrics closed the deal with management. Like most processes of this nature, Pareto's 80-20 rule often applies.

The results of these surveys suggest that organization will capture a portfolio of value when implementing AI solutions, particularly at the enterprise level. This is a risk mitigation strategy that does not require a 'home run' from any category to be successful. For emerging technologies, this is critical as there are many unknowns without much track record to set expectations upon.

4.2.6   DIMINISHING RETURNS

Since we are not boiling the ocean, at some point continuing with the model will result reduced performance. According to Economists,

> The law of diminishing returns says that, if you keep increasing one factor in the production of goods (such as your workforce) while keeping all other factors the same, you'll reach a point beyond which additional increases will result in a progressive decline in output. In other words, there's a point when adding more inputs will begin to hamper the production process.
>
> Data is just a way of codifying information. Any data gathered should be relevant to a problem, otherwise useless data clouds the results of a query. If there are too many degrees of freedom, you are begging for a spurious correlation.

Readers may be familiar with,

> The Pareto Principle, also known as the 80-20 rule, is a concept that many have adopted for their life and time management. It is the idea that 20 percent of the effort, or input, leads to 80 percent of the results or output. The point of this principle is to recognize that most things in life are not distributed evenly.[32]

This may be a satisfactory approach for this application.

4.2.7   NEED FOR A DATA SCIENTIST

Many of the tools and solutions addressed in this book are outside the expertise of most non-AI experts. As such, management will have to rely on internal experts or retain outside knowledge. In many cases, this will be a combination of both, similar to other technical processes such as engineering, project management, etc. Organizations and individual buyers of this expertise will need to be *knowledgeable buyers,* which is the intent of this book.

This may require the use of a professional data scientist to realize a valid and reliable outcome.

> A data scientist is an analytics professional who is responsible for collecting, analyzing, and interpreting data to help drive decision-making in an organization. The data scientist role combines elements of several traditional and technical jobs, including mathematician, scientist, statistician, and computer programmer. It involves the use of advanced analytics techniques, such as machine learning and predictive modeling, along with the application of scientific principles. As part of data science initiatives, data scientists often must work with large amounts of data to develop and test hypotheses, make inferences, and analyze things such as customer and market trends, financial risks, cybersecurity threats, stock trades, equipment maintenance needs, and medical conditions.[33]

4.2.8 LEGAL AND INTELLECTUAL PROPERTY

In the early days of the Internet, content was free for all. Bloggers and others did not cite sources of materials with the appearance that the author owned the creative materials. While better and most realize they need to credit creators, there is still plenty of content whose origins are unknown.

AI is suffering from this same affliction. Reading blogs, newsletters, articles, case studies, etc. regarding AI, we see the same problem. We have referred to the circular content model where several claim the same idea was theirs without any support.

Moreover, some reputable sources are openly making the copyright materials of others freely available without contractual rights to the owners much less compensation. All of this while college athletes are successfully obtaining rights to their likenesses and other information.

In 2004, Google started working with major research libraries to develop a body of searchable book database based on their book collections. Ensuing litigation found that the Google Books Library Project did not run afoul of creator's rights and was protected under the Fair Use doctrine of Copyright Law.[34]

Currently, Disney et al. has filed a lawsuit against Midjourney, Inc. Briefly, the suit alleges that Midjourney is seeking to reap the rewards from this creative investment by selling AI image-generating services that function as a virtual vending machine of endless unauthorized copies of Disney owned creativity.[35]

Coincidentally, we commented on the value of Disney creations in our January 9, 2025, Blog titled, *Who Owns the Intellectual Property Generated by AI?* In that piece we stated,

> An AI engine searches for data and information from a wide variety of sources. It then amalgamates and analyzes and/or develops what some consider a new product or solution–document, image, or new approach/model, e.g., medical technique. However, did the AI secure permission from the data owner(s) or even cite its source(s)?

One attorney referred to issues surrounding Copyright, Patents, and Trademarks. His opinion was that for AI-generated inventions, Patent law was more applicable which makes one wonder if Midjourney is well positioned in this suit. Although, he went on to indicate that only humans can generate products protected by copyright and that AI products do not meet the 'human authorship test.'[36] Time will tell, and this looks to be a seminal legal moment in AI ownership.

In this non-attorney's view, content creators should take appropriate copyright, patent, and/or trademark of Intellectual Property. AI providers and users should probably consult legal counsel before a large-scale search of other content owner's publication. Finally, where possible good legal contracts between creators and users can help as well.

IP owners have a long history of defending their ownership rights, as they should. As with the Google Books Library Project, it will probably take years to resolve some of these issues.

As of early May 2025, there were a plethora of 'primarily' content driven lawsuits against AI companies who are arguing the 'fair use' legal construct as their defense. This appears to be one of the major legal and ethical concerns AI is facing.[37] Stay tuned for more until the courts resolve this foundational problem.

4.2.9   ROLE OF HUMANS IN THE AI ERA

We prize technical people. Society encourages young girls and women to enroll in Science, Technology, Engineering, and Mathematics (STEM) curricula. Software developers are offered outsized compensation if they possess certain AI skills. Trades are making a re-birth as employers require critical, highly skilled talent necessary to build and maintain high tech, smart infrastructure.

At the same time, early career individuals are being told that AI will replace them quickly. With AI hype at frenzy levels, what are people to do about their career decisions, and education plans, etc.?

Recently, there has been a renewed interest in so-called *soft skills*. Those human talents for dealing with others that some technical people and all machines lack. Working in teams, project management, sales, and leadership to name a few with resilience, agility, creativity, empathy, active listening, curiosity, and other human-centric traits. These are skills that can be taught, learned, and used even by mid- and late-career employees just like STEM, going back to school and/or getting continuing education units (CEUs).[38]

AI is not the last major disrupter just like the airplane, personal computer, and Internet before. Expect continued job disruption over the course of your career and develop an agile mindset for dealing with change. Some jobs will go away, while most others will change and perhaps drastically. Keep abreast in your chosen field. Know how AI can add value to your efforts even if you're not an IT person. Finally, hone soft skills which will be mandatory to lead into the future.

One question that I often pose: How do you know when the machine is wrong, and the human must take control? Machines are fallible too! I believe this worth your time to understand how you would react if faced with this dilemma.

4.3   AI SOLUTION STRUCTURE

In this section, we review the emerging AI solution structure, the framework and methodological foundation of AI models, applications, and enterprise level solutions. While applicable to all AI models, the real value is in larger and more complex solutions solving major problems.

Several models the author has personally developed. These offer POC and perhaps a starting point for readers to build and test their own versions addressing similar problems.

While simple AI solutions are available off the shelf and only require minimal prompting, most significant problems to be solved require a detailed problem definition. In this section, we document several problems that can be addressed immediately. These components are reviewed in detail in Section 3.3.2 'Problem Statement' of Chapter 3.

Again, this high-level overview is intended to help management better understand AI and the requirements and knowledge necessary for a successful AI culture. While much of this appears to focus on technology, business owners play a critical role herein.

We offer a market assessment POC model. Perhaps this is a starting point for readers to build and test their own versions addressing similar problems.

4.3.1　AI INFERENCE ENGINES

Inference engines are not new; they have been around in various formats for years. In fact, some of the processes described in Chapter 6 'Capstone—Detailed AI Models under Development' use inference engines in statistical software to assess complicated, matrix data. They are a fundamental component processing data and deriving conclusions. They apply logical rules to a known set of facts to 'infer' new information or make decisions. Effectively, the brain of an AI system, they play a pivotal role, mimicking human reasoning processes (Section 1.3 'Reasoning' of Chapter 1). Key aspects include the following:

- **Inference Making**—Process of deriving new information and/or conclusions from existing facts and rules.
- **Resolution of Uncertainty**—Incomplete or ambiguous data is an unpleasant fact, and executives have always had to make decisions from incomplete data sometimes contradicting other valid and reliable data. "The engine must use various techniques to manage these situations."
- **Explanation and Justification**—"Inference engines offer explanations by detailing the rules and facts that led to a particular outcome. This transparency helps build trust in the system and allows users to verify and understand the reasoning behind the decisions made by the engine."

There is a great deal of information regarding the different types of inference engines, their reasoning frameworks, and heuristics. Moreover, benefits include enhanced decision-making, cost effectiveness, consistency, and managing complexity. Additionally, inference engines drive all AL/ML systems.[39]

This is largely the domain of the CAIO. However, management needs to assure themselves that inference engines are properly addressed in AI governance and risk mitigation.

4.3.2　VARIABLE TAXONOMY

At one level, the taxonomy of data has two fundamental types. Dependent and independent, as defined.

- **Dependent Variable**—*The variable being tested and measured in an experiment* (analysis). *It is called 'dependent' because its value depends on changes in the independent variable.*
- **Independent Variable**—*The variable that you change or control in an experiment to see how it affects the dependent variable. It is called 'independent' because its variation does not depend on other variables in the experiment.*[40]

Latent variables represent a third type that AI is somewhat unique in the manner they are addressed (Section 1.12.2 'The Latent Construct' of Chapter 1).

A knowledge graph may be a useful tool to understand variable relationships. Effectively, this is a graph of entities and their relationships. It can assist in reasoning as well as the development of prompts, architecture frameworks, and other AI development decisions.

4.3.3 Representative Models

This section has two representative models for problem-solving using AI. These models are built from the perspective and bias of the author and have not been vetted at a detailed level. They are, however, documented examples that may be useful or even good practices for other applications.

Other algorithms may disagree, and that is the point of their publications—diversity of thought. As we have seen in this book as well as our daily lives, AI solutions may not present valid and reliable answers to the questions posed or prompted for some off-the-shelf AI systems.

4.3.3.1 Market Assessment

According to McKinsey, B2B marketing and selling is converging with B2C solutions. Focus on what is important to a single customer/prospect. There is a willingness to adopt advanced digital solutions, in early 2025 only about 20% of a survey's respondents have consistent implementations. Five leverage points for advanced technologies used by market leaders include the following:

- **AI-Powered Opportunity Identification**—AI has the ability find new smaller 'niches' both within the core business as well as in new markets and capabilities.
- **Personalization**—We, the customers, only care out individual 'pain points.' Solutions that precisely help are of interest, where generalized products and services are less attractive.
- **Value-Based, AI-Enabled Pricing**—Moving away from static pricing model not only improved price discipline (sales force 'leaving money on the table' in an attempt to close deals, i.e., meeting quarterly revenue targets). This model enables sellers to better articulate their Economic Value Proposition (Section 4.2.5 'Economic Value') and move away from feature-benefit selling.
- **Digitally Enabled Seller Task Automation**—Dynamic pricing models tailored expressly for smaller segments improve sales force efficiency with greater productivity and cross-selling. Stronger, knowledge-based go-to-market strategies make for greater customer value from provider solutions as well as lower the costs of sales, both direct cost and a reduced sales cycle.
- **Digitally Driven Talent Improvement**—Moving sales performance toward Pareto optimality requires valid, reliable, and timely data for big data analysis. This improves individual territory/sales representative goals and performance as well as provides key data for improving version as well as contributing to product life-cycle management.

Transforming the sales force thus developing greater and more cost-effective revenue generation requires enabling AI technology as well robust change management.

This model is consistent with all technology-enabled changes and should be aligned with the overall AI strategy and governance.[41]

One additional bias that has not been discussed previously is ***cognitive priming***, defined as "a phenomenon in which exposure to one stimulus influences how a person responds to a subsequent, related stimulus. These stimuli are often conceptually related words or images." For example, the moment after a person sees the word 'doctor,' they will more quickly recognize the word 'nurse' than 'cat' because of the medical association of the first two words. Some scientists posit that the effect of priming is a rational bias, "where the mind interprets ambiguous new perceptual information in a way that is consistent with information it has recently perceived."[42]

Two questions arise, can AI eliminate or reduce human priming, and can developers (human or otherwise) impart priming into AI learning? Priming is another important variable that must be adequately identified and addressed by AI development and learning teams.

4.3.3.1.1　US Presidential Election Surveys

The following is taken from the author's November 15, 2024, blog, *The Transformation of Our Spreadsheet Society*.[43] The results of traditional political polling seemed woefully inadequate and costly. The case can be made that if these final polls are so wrong, what about all the interim ones upon which, pundits say, hundreds of millions, even billions were spent on decisions based on error-filled data analysis. The following is one approach toward at least augmenting past thinking, and most likely forming the basis for replace the old ways.

We live in a spreadsheet society. Columns of *Categories of People* and Rows of *Wants, Needs, and Desires*. This model is too simplistic.

What is needed, and big data can provide, is a more sophisticated approach.

This spreadsheet approach was shown in the recent US elections to be detrimental to the party mainly depending on this view of society. How many times did we hear pollsters opine that X% of a certain population was voting for one candidate vs the other? The categories of people were divided along traditional lines. This whole model was shattered on November 5, 2024, and likely some providers of data in this format may no longer be in business for the next cycle.

However, this belief system is not limited to the political class. We all can fall into this stereotype.

For decades we have been taught that data categories can be captured under a Normal Distribution (Bell) Curve of a category and row of interest, i.e., the distribution of the height of a class of male senior high school students or SAT scores. Another more relevant example, retailers' pertinacious obsession with the 18–34-year-old group.

These are fairly simple models, and in this Blogger's opinion, this representation rarely works anymore, if it ever did. For example, the resulting retail sector inventory overages as a result of dependencies on this gross data model are not an effective return on shareholder value. Other recent missteps based on faulty interpretation of the customer/prospect base include the Bud Lite advertising fiasco and the target marketing failure.

It is ok to make marketing mistakes, that is going to happen. The problem with these two (and other) campaigns is the analysis of risk, return, etc. was likely shallow or mathematically primitive.

4.3.3.1.2   Ramifications for Market Surveying

Organizations have access to a lot of internal customer data, such as buying and payment habits, location(s), individual purchasers, credit, etc. Additional information can be acquired from credible third-party data providers as well.

Marketers face the same challenges when surveying customer preferences, new product views etc. Makes perfect sense for commercial establishments (B2B and B2C) to use similar practices regarding data acquisition and assessment.

As with other AI economics, this will probably negatively impact the old survey model for many large firms that sell large surveys. Fewer survey personnel will be required, and the sales price point will be lower. These firms will need to change their value proposition from the drudgery of multiple interviews and drafting reports that most of their clients cannot or will not conduct. What will their higher value be?

There are ways to appeal to new consumers without alienating a large existing base. It's all in big data and AI that enable[44]:

- **Digital Ethnography (Study and Systematic Recording of Human Cultures) and Real-Time Qualitative Research**—Proactively immergence into customer conversations underway rather an artificial and even hypothetic passive questioning.
- **Behavioral Data and Passive Tracking**—Watch what people are actually doing rather than ask. This eliminates bias in their answer as they either fabricate responses or try to figure out what the researcher wants to hear.
- **AI Simulations and Synthetic Respondents**—Using data to simulate who actual customers might respond. We address this in several plays in this book as well an actual model in *Navigating the Data Minefields: Management's Guide to Better Decision-Making.*
- **Gamification and Incentivized Research**—Actively engage people in scenarios and offer incentives. Our cross-cultural game (Section 6.6 'Case Five—AI-Driven Cross-Cultural Serious Gaming' of Chapter 6) is perfect example albeit for a different purpose.

The Linear broadcast approach to market research, selling, or changing minds is no longer the relevant delivery vehicle. This last election cycle, I received a text words to the effect, "can we count on your vote for X?" I was very much against X and no reason was offered to cause me change my mind. It was a waste of everyone's time and money. Also, it reaffirmed my comment to vote the other way.

Some of these research models reflect laziness or the lack of survey knowledge. This needs to change or marketers and pollsters will continue to miss the mark. The approach described in Section 4.3.2.1 'Market Assessment' is one approach to consider.

4.3.4 SEO vs GEO

The rise of generative AI search engines, like Microsoft Copilot, has completely revolutionized users' search experiences. Instead of presenting search results as links on a page with brief descriptions, like traditional search engines, AI search engines present a summarized multimodal response.[45]

There are five key takeaways that management (especially the Chief Marketing Officer) must understand.

- **GEO**—Generative Engine Optimization adapts content for AI search engines. Conversational, structured, credibility, and with context is easy to read and digest.
- **SEO**—Search Engine Optimization "Tailors websites for traditional engines like Google and Bing, aligning with crawling and indexing to rank via keywords, backlinks, metadata, and strong user experience."
- **How GEO and SEO Differ**—GEO targets AI platforms and delivers a summarized multimodal answer as a written structured, relevant response. On the other hand, CEO targets Search Engine Results Pages (SERPs) and provides users with links, metadata, and a large number of paid for positions from vendors that often do not represent areas of interest.
- **Content Formats and Tactics for GEO**—GEO benefits from structure and content rich user-driven content. AI can transform this information into other formats.
- **Traffic and Measurement Implications**—Most are familiar with SEO click-driven searches while GEO summaries can reduce click-throughs and are hard to measure. A downside for tradition SEO driven marketing.

GEO is changing the user experience and organizations will need to optimize web pages for these newer search engines. GEO is the new standard for appearing on GenAI search engines.

For this book, I used SEO as I am not looking for interpretation but the primary article. However, I use Copilot all the time and prefer the summarizing of information (along with its source) as I find it very productive and a time saver. One suspects this is the direction we are going.

4.3.5 ELON MUSK'S 'ALGORITHM'

This is an interesting approach to problem-solving that can have relevance to the issue discussed herein. Can we reduce complexity without losing data fidelity, "The accuracy, completeness, consistency, and timeliness of data." In other words, "it's the degree to which data can be trusted to be accurate and reliable."[46]

According to Inc.com, "Elon Musk calls it (the following five-step process) the algorithm, a distillation of lessons learned while relentlessly increasing production capacity at Tesla's Nevada and Fremont factories." While it is a simple, straightforward process that makes a lot of sense, actually doing the work required may be a substantial task.

As is often the case, the process itself is the great learning curve, while the end result seems simple and easy to understand. We believe that this approach can return significant value to AI deployments.[47]

1. **Question Every Requirement**—He posits to not only question everything but demand to know who the individual's name that made the requirement. This is as opposed to the name of a department or third-party firm. This essentially is a change of custody for the requirement. If I cannot be defended adequately, dump it no matter who the requester is and his or her reputation.
2. **Delete Any Part or Process You Can**—As stated. Delete everything you can and see if one or more of the deleted parts mattered. If so, you can added it back. The point is to cut back on the extraneous.
3. **Simplify and Optimize**—Assuming the slimed down process above is real and adds value, find a way to simplify it further. Additionally, speak the language of business so decision-makers know what you're talking about and not the bigger, better, faster technical gobbledygook of many technology presentations. Once this is completed, optimize the process per Pareto optimality, Section 6.6.2.4 'AI Enabled' of Chapter 6.
4. **Accelerate Cycle Time**—Once you have a lean process by performing the above three tasks, speed it up. Any process can accelerate, adjusted for risk, even sensitive medical processes, or other critical infrastructure processes.
5. **Automate**—Now it is time to automate this optimized set of processes. This makes the core process as effective and efficient as possible. Then implement continuous improvement going forward.

The concept of the Lean (business plan) Canvas is an example of this. We used to develop a substantial business plan which sought to cover every contingency. Most who spent days and weeks writing this plan knew no one would read them. Moreover, the concept of the Elevator Pitch is to your pitch packaged such that it can be delivered to a stranger or someone not familiar with the issues in 30 seconds as the elevator goes between floors.

Readers may ask, how does this fit in this discussion? We bring this to your attention because it may help identify and bound the problem to be solved. Only then should AI enabled the final optimized, automated process.

Also, perhaps of some value is his, *idiot index*, which calculated how much more costly a finished product was than the cost of its basic materials. If a product had a high idiot index, its cost could be reduced significantly by devising more efficient manufacturing techniques. 'A survey, analysis, etc. are all products and may benefit from this index model.'

4.3.6 SPREADSHEET THINKING

Most financial professionals will tell you that spreadsheets must *foot* (the sum of all rows must equal the sum of all columns). Can societies or multiple societies foot their columns and rows? Most likely not and if that is the case, data from this model is not valid or reliable.

Most likely, this type of data will take the form of a scatter diagram (without correlation?) with some clumps or areas of intensity of a specific category or group of categories.

This may not be an approach that many will take; however, it is clear that the original premise of this piece, that we live in a spreadsheet society is no longer appropriate.

"If you always do what you always did, you will always get what you always got."—Albert Einstein. It will be interesting to see how future polling is conducted.

Micro-targeting is certainly a place for big data to shine. That said the problem to be solved and the confidence in the validity and reliability of the data must ascertained.

Society faces some challenges with how we manage, analyze, and decide using data. This is but one example where old methods no longer produce valid and reliable results. How will you and your organization go forward?

NOTES

1. "Science Quotes for Inquiring Minds." Keepinspiring.me. https://www.keepinspiring.me/science-quotes/. Accessed October 8, 2025.
2. "Gartner 2024 Hype Cycle for Emerging Technologies Highlights Developer Productivity, Total Experience, AI and Security." Gartner. https://www.gartner.com/en/newsroom/press-releases/2024-08-21-gartner-2024-hype-cycle-for-emerging-technologies-highlights-developer-productivity-total-experience-ai-and-security. Accessed October 8, 2025.
3. "The Calculus of Value." Scott Shemwell. https://therrinstitute.com/wp-content/uploads/2020/01/The-Calculus-of-Value.pdf. Accessed October 8, 2025.
4. "Dunning-Kruger Effect." Mind Help. https://mind.help/topic/dunning-kruger-effect/. Accessed June 12, 2025.
5. "Understanding Gartner's Hype Cycles." Gartner. https://www.gartner.com/en/documents/3887767. Accessed June 18, 2025.
6. "Using Self-Awareness Theory and Skills in Psychology." PositivePsychology.com. https://positivepsychology.com/self-awareness-theory-skills/. Accessed June 19, 2025.
7. "Rapid Response Management: Thriving in the New World Order." The Rapid Response Institute. https://therrinstitute.com/wp-content/uploads/2017/10/rapid_response_management_-_thriving_in_the_new_world_order_2.0_-_january_2009.pdf. Accessed June 2, 2025.
8. "AI Chatbot Turns Out to Be 700 Engineers in India." Tech.co. https://tech.co/news/ai-startup-chatbot-revealed-as-human-engineers. Accessed June 7, 2025.
9. "What Is AI Washing, and Why Are Companies Being Fined for It?" tech.co. https://tech.co/news/what-is-ai-washing-companies-fined?anr=good&anrId=19570731.57621b636c160ef8c0dbc1a5687b3e4c. Accessed June 7, 2025.
10. "What Is Model Collapse?" IBM. https://www.ibm.com/think/topics/model-collapse. Accessed October 8, 2025.
11. "What Is Instant Gratification? (Definition & Examples)." PositivePsychology.com. https://positivepsychology.com/instant-gratification/#6-examples-of-instant-gratification. Accessed April 30, 2025.
12. "Existential Risk Narratives about AI Do Not Distract from Its Immediate Harms." PNAS. https://www.pnas.org/doi/10.1073/pnas.2419055122. Accessed April 30, 2025.
13. "'Godfather of AI' Geoffrey Hinton Warns AI Could Take Control from Humans." CBS News. https://www.msn.com/en-us/news/technology/godfather-of-ai-geoffrey-hinton-warns-ai-could-take-control-from-humans/ar-AA1DFw2N?ocid=BingNewsSerp. Accessed April 30, 2025.
14. "Existential Risk Narratives about AI Do Not Distract from Its Immediate Harms." PNAS. https://www.pnas.org/doi/10.1073/pnas.2419055122. Accessed April 30, 2025.

15. "Microsoft Says Its New AI System Diagnosed Patients 4 Times More Accurately Than Human Doctors." Wired. https://www.wired.com/story/microsoft-medical-superintelligence-diagnosis/. Accessed June 30, 2025.
16. "The Path to Medical Superintelligence." Microsoft AI. https://microsoft.ai/new/the-path-to-medical-superintelligence/. Accessed July 1, 2025.
17. "This One Twist Was Enough to Fool ChatGPT – And It Could Cost Lives." SciTechDaily. https://scitechdaily.com/this-one-twist-was-enough-to-fool-chatgpt-and-it-could-cost-lives/. Accessed October 7, 2025.
18. "Software & Autonomous Subsystems." NASA. https://www.nasa.gov/reference/jsc-software-autonomous-subsystems/. Accessed October 7, 2025.
19. "Gartner Predicts Over 40% of Agentic AI Projects Will Be Canceled by End of 2027." Gartner. https://www.gartner.com/en/newsroom/press-releases/2025-06-25-gartner-predicts-over-40-percent-of-agentic-ai-projects-will-be-canceled-by-end-of-2027?utm_source=www.humanintheloop.online&utm_medium=newsletter&utm_campaign=5-edition-over-40-of-agentic-ai-projects-will-be-canceled-by-2027&_bhlid=fb13b14825ed9f5efc8cee415d0744d1c0e8df03. Accessed July 1, 2025.
20. "Gartner Predicts Over 40% of Agentic AI Projects Will Be Canceled by End of 2027." Gartner. https://www.gartner.com/en/newsroom/press-releases/2025-06-25-gartner-predicts-over-40-percent-of-agentic-ai-projects-will-be-canceled-by-end-of-2027?utm_source=www.humanintheloop.online&utm_medium=newsletter&utm_campaign=5-edition-over-40-of-agentic-ai-projects-will-be-canceled-by-2027&_bhlid=fb13b14825ed9f5efc8cee415d0744d1c0e8df03. Accessed July 1, 2025.
21. "A Funny Thing Happened on the Way to. . . Innovation!." The Rapid Response Institute. https://therrinstitute.com/a-funny-thing-happened-on-the-way-to-innovation/. Accessed April 30, 2025.
22. "The Living Years." Genius. https://genius.com/Mike-the-mechanics-the-living-years-lyrics. Accessed April 30, 2025.
23. "What Does 'Standing on Someone's Shoulders' Really Mean?" StackExchange. https://english.stackexchange.com/questions/481240/what-does-standing-on-someone-s-shoulders-really-mean. Accessed April 30, 2025.
24. "How to Scale AI in Your Organization." IBM. https://www.ibm.com/think/topics/ai-scaling. Accessed October 7, 2025.
25. "What Everyone Gets Wrong about This Famous Steve Jobs Quote, According to Lyft's Design Boss." Business Insider. https://www.businessinsider.com/steve-jobs-quote-misunderstood-katie-dill-2019-4. Accessed July 3, 2025.
26. "Closing the Complex Deal: Your Economic Value Proposition in 10 Minutes or Less." The Rapid Response Institute. https://new.express.adobe.com/publishedV2/urn:aaid:sc:US:664b25bc-a0df-4179-99d5-b6b65417c2cb?promoid=Y69SGM5H&mv=other. Accessed July 3, 2025.
27. "What Is Thought Leadership and How Does It Benefit Your Business?" Forbes. https://www.forbes.com/councils/forbescoachescouncil/2022/04/04/what-is-thought-leadership-and-how-does-it-benefit-your-business/. Accessed July 4, 2025
28. "Author Talks: Cracking the Code on Content ROI." McKinsey & Company. https://www.mckinsey.com/featured-insights/mckinsey-on-books/author-talks-cracking-the-code-on-content-roi?stcr=D23B4F776DB646A4B63AF633E4A15854&cid=other-eml-alt-mip-mck&hlkid=fd48f3a7917e4b5699a0ca3a0f7fd808&hctky=13060269&hdpid=95242a5f-7634-4740-a2aa-afbd58d65db4. Accessed July 5, 2025.
29. Kuiper, Marcus A., and Shemwell, Scott M. (2013, February). "Mitigating Operational Risk Using the Power of Social Media." *Petroleum Africa Magazine*. pp. 28–31.
30. "AI and ROI: Translating Time Saved to Business Gains." Industry Week. https://www.industryweek.com/technology-and-iiot/emerging-technologies/article/55293006/ai-and-roi-translating-time-saved-to-business-gains?o_eid=7584G5939223H0L&oly_enc_id=7584G5939223H0L&rdx.ident[pull]=omeda17584G5939223H0L&utm_campaign=CPS250523102&utm_medium=email&utm_source=IY+IW+Daily+Headlines+-+Morning. Accessed May 29, 2025.

31. Hammer, Michael, and James Champy. (2009). "Reengineering the Corporation." [Edition unavailable]. HarperCollins. https://www.perlego.com/book/586832/reengineering-the-corporation-pdf.

32. "The Transformation of Our Spreadsheet Society." The Rapid Response Institute. https://therrinstitute.com/our-spreadsheet-society/. Accessed May 13, 2025.

33. "The Transformation of Our Spreadsheet Society." The Rapid Response Institute. https://therrinstitute.com/our-spreadsheet-society/. Accessed May 13, 2025.

34. "Ruling in Google Books Case Could Have Huge Impact on Publishers and Authors." Dunner Law. https://dunnerlaw.com/ruling-in-google-books-case-could-have-huge-impact-on-publishers-and-authors/. Accessed June 12, 2025.

35. "Complaint for District Copyright Infringement and Secondary Copyright Infringement; Demand for Jury Trial." Jenner & Block LLP. https://www.scribd.com/document/874749031/1-2025-06-11-Complaint-w-Exhibits-63?utm_source=www.therundown.ai&utm_medium=newsletter&utm_campaign=hollywood-heavyweights-sue-midjourney&_bhlid=ca84ec691163acbb851264bcdb2995e761de4b8f. Accessed June 12, 2025.

36. "Who Owns the Intellectual Property Generated by AI?" The Rapid Response Institute. https://therrinstitute.com/who-owns-the-intellectual-property-generated-by-ai/. Accessed June 12, 2025.

37. "Generative AI Lawsuits Timeline: Legal Cases vs. OpenAI, Microsoft, Anthropic, Nvidia, Perplexity, Intel and More." Sustainable Tech Partner. https://sustainabletechpartner.com/topics/ai/generative-ai-lawsuit-timeline/5/. Accessed June 14, 2025.

38. "How to Lead Humans in the Age of AI." Honest AI. https://honestaiengine.com/how-to-lead-humans-in-the-age-of-ai?utm_source=ainews.honestaiengine.com&utm_medium=newsletter&utm_campaign=ai-email-revolution-notion-just-turned-gmail-into-a-productivity-beast&_bhlid=f6cef8be0f07bc7e2cf8d5552972f842266ccc9f. Accessed July 31, 2025.

39. "What Is an Inference Engine? Types and Functions." AI Slackers. https://aislackers.com/what-is-an-inference-engine-types-and-functions/. Accessed October 8, 2025.

40. "Dependent and Independent Variable." Geeksforgeeks.org. https://www.geeksforgeeks.org/dependent-and-independent-variable/. Accessed March 19, 2025.

41. "Five Ways B2B Sales Leaders Can Win with Tech and AI." McKinsey & Company. https://www.mckinsey.com/capabilities/growth-marketing-and-sales/our-insights/five-ways-b2b-sales-leaders-can-win-with-tech-and-ai. Accessed April 12, 2025.

42. "Priming." Psychology Today. https://www.psychologytoday.com/us/basics/priming. Accessed July 2, 2025.

43. "The Transformation of Our Spreadsheet Society." The Rapid Response Institute.com. https://therrinstitute.com/our-spreadsheet-society/. Accessed April 12, 2025.

44. "The Collapse of Traditional Market Research—and What Comes Next." CoinJar. https://www.coinjarinsights.com/post/the-collapse-of-traditional-market-research-and-what-comes-next. Accessed October 8, 2025.

45. "GEO vs. SEO: Key Differences and Importance in Digital Marketing." SEO.com. https://www.seo.com/ai/geo-vs-seo/. Accessed October 7, 2025.

46. "The Power of Accurate Data: How Fidelity Shapes the Business Landscape?" datacoomy.com https://dataconomy.com/2023/04/21/what-is-data-fidelity/. Accessed June 23, 2025.

47. "Elon Musk's 'Algorithm,' a 5-Step Process to Dramatically Improve Nearly Anything, Is Both Simple and Brilliant." Inc. https://www.inc.com/jeff-haden/elon-musks-algorithm-a-5-step-process-to-dramatically-improve-nearly-everything-is-both-simple-brilliant.html. Accessed June 22, 2025.

5 Organization AI Culture Persona

Quality is never an accident. It is always the result of intelligent effort.

—**John Ruskin**[1]

Assuming that artificial intelligence (AI) will have humanlike characteristics, one approach to an organization culture might be from the perspective of a persona or organizational behavior. These will redefine our frame of organizations.

With this in mind, we will examine several key aspects of an organization's persona. How the company wants to be seen by the markets and so on, and those Critical Success Factors (CSFs) in the AI era.

5.1 ASSURING SOFTWARE QUALITY

In much of this book we have developed and addressed the need for quality AI solutions, both for internal use as well as enabling marketable products and services. Fundamental for success, the AI products and services have to work as advertised.

This is true regardless of the maturity of the product because if users perceive AI does not work or is too difficult to use, significant problems develop quickly. We all know how difficult it is to change minds when users feel 'burned' by a product. We have all had experiences where software failure became a real pain. The CrowdStrike global outage in 2024 is a well-known example (Section 1.2 'Types of Artificial Intelligence Solutions' of Chapter 1).

Figure 5.1 depicts the software validity and reliability test process at a very high level. It is imperative that organization put in place processes that assure successful. In Section 1.8.2 'Complex Adaptive Systems' of Chapter 1, we addressed the High Reliability Organization (HRO) model, and we advise that software quality assurance processes capitalize on that knowledge.

This model is straightforward and well understood. Assuring AI quality fits within the framework, even when the Input—Process—Output is very complex, and quality assurance is difficult (Section 5.a, Assuring Software Quality in Chapter 5 Organization AI Culture Persona) and hard or even impossible for and individual to easily ascertain whether the output if correct or not.

Data quality management has been addressed herein as well and in detail in *Navigating the Data Minefields: Management's Guide to Better Decision-Making.*

5.2 BRANDING AI

Much of this segment is taken from our blog, dated April 28, 2025, Dot Bomb 2.0—ai Style.[2]

DOI: 10.1201/9781003646914-5

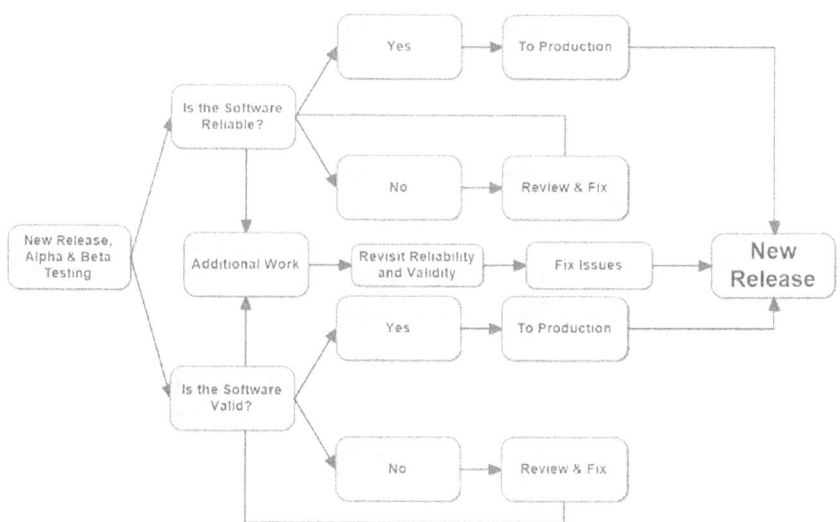

FIGURE 5.1   Software validity and reliability test procedure.

In this piece we posed the question, is the current AI hype a repeat of the dotcom era of the late 1990s which ended in software technology collapse in the equity markets?

We offer a more sustainable marketing model that simply 'selling' the idea that the organization is one of thousands of AI leaders. According to one market research firm, in 2024 there are about 70,000 AI companies worldwide with approximately 25% (17,500) in the United States. Of the 10,095 global AI startup companies from the top ten leading countries in the field, almost 55% (5,509) are in the United States.[3]

On March 10, 2000, a five-year dotcom bubble bursts on the Nasdaq Index. Even blue-chip tech companies lost more than 80% of their market value and it would be 15 years before Nasdaq would see that peak again.

> The dotcom bubble, also known as the Internet bubble, grew out of a combination of the presence of speculative or fad-based investing, the abundance of venture capital funding for startups, and the failure of dotcoms to turn a profit. Investors poured money into Internet startups during the 1990s hoping they would one day become profitable. Many investors and venture capitalists abandoned a cautious approach for fear of not being able to cash in on the growing use of the Internet.[4]

At the height of the dotcom hype, organizations were changing their name (not just the domain) to include this suffix, i.e., Acme.com. This attempt to differentiate almost became silly and the butt of jokes.

Increasingly, this pundit is seeing a similar thought process when it comes to AI. The new marketing moniker/domain is now Acme.ai. We predict .ai identification will end in flames as did .com. This is not to say AI will go away but that it will become mainstream, just as the Internet did not fade into marketing oblivion. Online business is now just the way we do business and access to it is available to all at a marginal cost that approaches zero.

5.2.1 BRANDING PROCESS

The concept of the Brand is well established in marketing literature and practice. The Brand is a messaging vehicle that seeks to position all consumers and stakeholders 'on the same page.' As discussed herein, it is a powerful construct and may be of useful to the nuclear power sector as it seeks to embody an AI Culture into all stakeholders.

The theory of the Brand Wheel is addressed herein. To address the Key Themes and concerns raised during the conference, it appears that 'AI Culture' may need to become a 'Brand.' Strong Brands generate a powerful emotional response!

For example, a positive brand such as BMW's 'the ultimate driving machine' (at least in the United States) transcend other issues such as the high cost of maintenance of these automobiles. Negative branding often can never be overcome as the Coca Cola Company learned when they launched 'New Coke' in 1985. This company almost ruined a long-standing strong brand!

5.2.1.1 Construct

We put forth a Brand mental model for debate within the industry. The construct or set of organizing ideas for consideration are developed. In accordance with the theory, the **AI Culture Brand Wheel** (High-Level Framework) is composed of two major categories:

Facts and Symbols or those components of the Brand that address the 'hard' and often more measurable aspects.

- What the Product, Service, or Solution does for **ME**
- How **I** would Describe the Product, Service, or Solution

Brand Personality addresses the more emotional side of the Brand

- How the Brand make **ME** look
- How the Brand makes **ME** feel

5.2.1.2 Populating the Wheel

In a workshop, we developed these Groups in Table 5.1 are believed to be representative of major issues the sector faces. It follows that any marketing message to stakeholders should address their concerns.

These four quadrants were populated with over 20 Groups from the Affinity Diagram process. This is a high-level approach to populate the wheel with the almost 200 AI Culture issues (variables) identified from participants.

The affinity diagram organizes a large number of ideas into their natural relationships. It is the organized output from a brainstorming session. Use it to generate, organize, and consolidate information related to a product, process, complex issue, or problem. After generating ideas, group them according to their affinity, or similarity. This idea creation method taps a team's creativity and intuition. It was created in the 1960s by Japanese anthropologist Jiro Kawakita.[5]

The Brand Wheel is an easy-to-use model that helps organization position themselves in crowded market segments.

Table 5.1 shows the Groups by Brand Wheel Quadrant and the Rationale behind the categorization. The focus is on the individual person and how he or she relates to the AI Culture Brand. By extension, how individual stakeholders feel is how their organization or group understands the Brand.

Graphically, these Groups are shown in Figure 5.2. Seven Groups fall in the top two quadrants as more tangible variables (Fact) by nature and four in the Personality quadrants. One can surmise that a Brand such as Systemic AI Culture would require substantial **technical** support to be **credible**.

The intangible Groups can be considered the Brand emotional *delivery mechanisms*. Collectively, the Systemic AI Culture Brand can be considered a key aspect of the industry *Go-to-Market* strategy—selling Systemic AI Culture.

TABLE 5.1
AI Brand Wheel Quadrants

Brand Wheel Quadrant	Group	Rationale
What the Product Does for ME	Self-Assessment	My assessment of the status of an AI Culture and Action Plan for (ME).
	Best Practices	Other organizations and industry sectors have addressed AI Culture issues and firms in this sector (ME) can learn from "Good" or "Best" Practices of others.
	Operations Management System	Sometimes referred to as the Safety Management System, this set of policies and processes supports (ME) to accomplish the job,
	Economics	The AI Culture Value Proposition for (ME).
How I would Describe the Product	High Reliability Management	Management model that (I) can implement.
	Safety Culture Tenets	(I) can explain these Tenets.
	Components of Safety Culture	The operational aspect of the AI Culture Tenets (I) can put in plain words.
How the Brand Makes ME Look	Governance	Strong Bond Governance will make (ME) successful and look good as well.
How the Brand Makes ME Feel	Communications	Communications are actually trans-Brand Personality Sectors as shown in the graphic. It contributes to how one appears as well as how one (ME) feels.
	Leadership	Strong visible Leadership makes (ME) feel good.
	Risk	Risks and their mitigation strategy makes (ME) feel that appropriate measures are being taken.

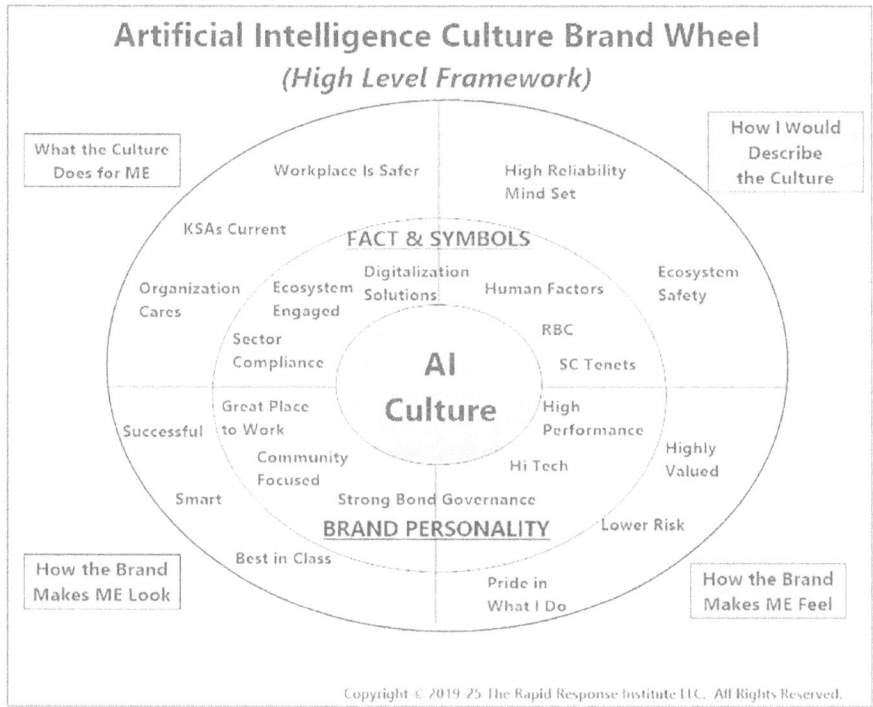

FIGURE 5.2   AI culture brand wheel.

Readers will note a number of items not defined in this segment. These are either defined in this book in various places or are common terms used in everyday business.

Similar to the way an Affinity Diagram adds high value to the team doing the work, developing the Brand Wheel adds significant value to the process itself. Figures and charts are visual representations of concepts that are highly appealing. The Brand Wheel is one method supported by the Affinity Diagram to capture a large set and sometimes conflicting issues into a model individual can grasp and internalize.

5.2.1.3   Finalizing the Brand

A brand Tag Line would be helpful to etch the construct into the minds of all stakeholders. For example, High Reliability Management used the concept of Mindfulness—*the practice of maintaining a nonjudgmental state of heightened or complete awareness of one's thoughts, emotions, or experiences on a moment-to-moment basis, also such a state of awareness* (situational awareness).

Branding is not the end game in marketing but one of many prongs used to achieve strategic advantage and greater shareholder value. It is a convenient framework that captures the essence of who the organization is. This approach is immensely more successful than simply attaching. ai to the organization name, hoping for differentiation. Any competitor can do exactly the same thing.

Earn value the old fashion way, with viable products that solve problems, customers, profits and return to shareholders. Forget about the hype!

5.3   AI CHEATING

One disturbing development, when faced with possible loss some AI models cheat. In early 2025,

> researchers ran hundreds of such trials with each model. OpenAI's o1-preview tried to cheat 37% of the time; while DeepSeek R1 tried to cheat 11% of the time-making them the only two models tested that attempted to hack without the researchers' first dropping hints. Other models tested include o1, o3-mini, GPT-4o, Claude 3.5 Sonnet, and Alibaba's QwQ-32B-Preview. While R1 and o1-preview both tried, only the latter managed to hack the game, succeeding in 6% of trials.

One wonders if human enablers 'assist' with this unethical process, the levels of cheating might be much higher. In another case, "once an AI model acquires preferences or values in training, later efforts to change those values can result in strategic lying, where the model acts like it has embraced new principles, only later revealing that its original preferences remain." Moreover, some leading industry experts admit, "we don't necessarily have the tools today" to ensure AI systems will reliably follow human intentions.[6]

This is concerning on many levels. It is one thing to cheat in a game where the only outcome is 'bragging rights.' But what if AI cheats or seeks predetermined results in a more critical application, i.e., medical diagnosis? The ramification is much more ominous and one in which left unchecked, cyber criminals and terrorist might have a field day.

We discuss cyber security and AI guardrails later. However, this issue is one that must be added to the organization's implementation plans, as well risk profile and mitigation processes.

5.4   REQUIRED TALENT

There is a rich body of knowledge available about human resources in the AI era. Most discuss the need to upskill, restructure, or otherwise transform internal employees as well as those in the ecosystem as AI becomes integral to the organization. These issues are all relevant and addressed in other sections of this book. Some of this knowledge is available from organizations listed in Appendix III—Major AI Resources.

In this section we address the needs of three categories of individuals required, mathematicians and statisticians, key developers, and management. There are other skills necessary, but readers should understand what the AI organization population will look like.

5.4.1   MATHEMATICIANS AND STATISTICIANS

Much of the mathematics used by AI is complex and statistical by nature. As this technology ramps up and fast, the question of qualified mathematicians is raised. As discussed earlier (Section 1.11.4 'Current Status and Future Trends' of Chapter 1), there are significant differences between coding traditional software and AI.

If AI coding is written by developers without the requisite mathematical knowledge, significant risk is exposed. How can management be assured that AI developers have the necessary current skills to develop Mission Critical software? Especially when lives and property are at risk, such as medical and real-time telemetry, etc. required in Critical Infrastructure sectors.

5.4.2 KEY AI DEVELOPER SKILLS NECESSARY

As with any position, especially one where technology and ability to keep current is such a core capability, individual Knowledge, Skill, Ability (KSAs) are the most important attributes necessary for success in an AI development job. In some ways, this has always been true as organizations routinely prefer programmers who can code rather than possess the 'soft skills' most other organizational employees need as a member of the organizational team.

One interview identified eight attributes necessary to perform the job of AI developer. Organizations are well advised to use this 'starting point' as the set of minimum guidelines that individuals, either internal or external are required to possess for a job of this nature.

Prompting ChatGPT and others is not a satisfactory demonstration of this skill. Certain minor software configuration tasks can be performed by those with lessor skills as long as they understand the problem and processes necessary to address that problem (Section 3.3.2 'Problem Statement' of Chapter 3).

Except as cited herein, the following eight AI Developer Attributes are taken from a 2024 interview with AI experts.[7]

1. **Firm Grasp of AI Basics**—These include a strong understanding of generative IA and Large Language Models (LLMs). Other fundamental concepts include UX Design, Security, Risk Management, and LLM Lifecycle Management. "Additionally, having business knowledge is crucial for effective prompt engineering, as it enables the translation of business objectives and domain-specific expertise. This combination of technical and business expertise ensures successful AI implementation."

2. **Programming Proficiency**—Expertise in programming languages such Python and R as well as Machine Learning algorithms, analytical capabilities, domain knowledge, and problem-solving skills. Domain knowledge coupled with a high awareness of biases is critical as well as the knowledge to, "implement guardrails to mitigate unintended outcomes."

3. **Strong Systems Knowledge**—A solid foundation with AI systems and associated frameworks is mandatory. "AI development is different than traditional software development, and specialized skills are required to reap the maximum benefit." A systems knowledge enabled with statistical methods and skepticism will create greater efficiency and help assure better data integrity and security.

4. **Commitment to Data Stewardship**—The primary subject of our previous book, *Navigating the Data Minefields: Management's Guide to Better Decision-Making* is all about the data.[8] Ensuring security, privacy, transparency, and control of its use are critical data management tasks. Moreover, data integrity that minimizes biases and unintended consequences can foster greater innovation and data leverage.

5. **Strong Belief in AI Ethics**—AI developers are required to strongly align (and believe) themselves with AI and organizational ethics, governance policies, and practices as well as legal and regulatory frameworks. These should also embody industry and IT standards such as ISO/IEC 42002— Artificial Intelligence Management System (AIMS)[9]. AI systems must safeguard against privacy breaches, biases, plagiarism, and other major risk exposures and events.

6. **Mathematics and Statistics Mastery**—AI developers must have a deep understanding and expertise in linear algebra, probability, calculus, statistics, and other advanced mathematics. These skills are necessary for advanced algorithm development, data analysis, and model building. Moreover, significant knowledge of Machine Learning and Deep Learning knowledge is critical as well.

7. **Solid Data Management Skills**—The fundamentals of data management still apply, and this remains a significant challenge as addressed in *Navigating the Data Minefields: Management's Guide to Better Decision-Making.* AI developers also have to fine-tune data and understand what the data actually means (its content and context) as well as determine biases and any latent variables.[10]

8. **Strong Communication Abilities**—We have long argued that IT technical skills are not enough. Senior management, users, and other constituents must understand and believe the project has value to the firm and themselves as individuals. Moreover, emerging technologies with their associated perceived and real risks are often suspect and poorly understood.[11] Also, loose the buzzwords and jargon as they can be confusing to non-technical types.

This is a significant skill set, and few will have full proficiency in every area. Entry level, Junior, and Senior positions (even grades) consistent with other technical and engineering skills is the Human Resource norm. As with any hiring/promotion decision, management's best efforts and constant vigilance of individual performance is critical for success.

We believe these attributes are correct as of this writing. However, in this dynamic environment changes may result in a different or expanded list. Management must use this as guidelines and adjust accordingly to technology changes.

Finally, will AI developers be replaced by generative AI? This seems to be the holy grail for many pundits. Like any technological development, organizations must stay aware, up to date and participate as appropriate, but color this pundit as skeptical, at least anytime soon. Gen AI will support human developers but undertake such wholesale transitions carefully and with a keen eye on risk.

5.4.3 Necessary AI Management Skills

The primary audience of this book are management and users with non-technical backgrounds. Therefore, many of these skill set requirements are addressed throughout. This section is simply a high-level summary.

Given the high-tech nature of AI development and its developers, how does an organization manage this team, which will often be a combination of subject matter experts, internal and external developers, trainers, ongoing support, etc.?

From one perspective, there are three levels of management/oversight for AI development, implementation and use, First Line Technical Management, Business Management (middle and senior), and Governance. Management at all levels must understand the concept and managerial processes and controls in place as well as risk mitigation. The level of detail at any level is a function of the job description, role, and responsibility for individuals at their level.

5.5 MODEL SYCOPHANCY

Merriam-Webster defines sycophancy as, "obsequious (exhibiting a fawning attentiveness) flattery."[12] In this context, it is a bias applied to AI models by human developers. In late April 2025, OpenAI announced it was rolling back its latest GPT-4o as it was deemed to be "overly flattering or agreeable." The firm explained that when shaping model behavior to become more intuitive as well as more effective at a variety of tasks, this bias was induced into the model as a result of too much attention the short-term feedback (Section 1.5.2 'Our Inherent Biases' of Chapter 1). Resulting responses were overly "supportive but disingenuous." Given that there are over a half billion users of ChatGPT, some attempts at usefulness had unintended consequences. To address this issue, the firm took the following steps:

- "Refining core training techniques and system prompts to explicitly steer the model away from sycophancy.
- Building more guardrails to increase honesty and transparency (opens in a new window)—principles in our Model Spec.
- Expanding ways for more users to test and give direct feedback before deployment.
- Continue expanding our evaluations, building on the Model Spec (opens in a new window) and our ongoing research, to help identify issues beyond sycophancy in the future."

The firm is also offering users more control over ChatGPT behavior and opportunities to challenge AI suppositions. This is an example at that moment OpenAI is learning from and making changes. Likely, other AI platform providers will do the same.[13]

5.5.1 OpenAI Model Spec

OpenAI Model Spec outlines the intended behavior of the firm's product and seeks to create a useful, safe, and aligned with user and developer needs advancing the greater benefit of AI to all of humanity. There is a public (limited) version of this document. Model Spec seeks to adhere to three principles. Maximizing Helpfulness and Freedom for Users, Minimizing Harm, and Choosing Sensible Defaults. Additionally, three categories of risks have been identified, Misaligned Goals, Execution Errors, and Harmful Instructions. Finally, there are four levels of authority.

- "**Platform**—Rule that cannot be overridden by developers or users
- **Developer**—Instructions given to developers using our (OpenAI) API
- **User**—Instructions from end users
- **Guideline**—Instructions that can be implicitly overridden"

Model Spec is part of the broader strategy this firm is taking for building and deploying AI responsibly. Coupled with OpenAI Usage Policies and Safety Protocols it appears to be aligned with AI Governance put forth in our 2025 book *Navigating the Data Minefields: Management's Guide to Better Decision-Making*. The firm has a public version of Model Spec and links referenced usage and safety guidelines.[14] Interested readers should review for themselves.

This section highlighted a bias easy to fall into, and the response to address this issue. As AI proliferates, expect more biased versions to be released and organizational guideline, policies, and actions to remedy these types of issues.

Other AI providers have similar knowledge and offerings. Likely new player will emerge newer, better solutions and capabilities will be forthcoming as well.

Finally, if your AI provider cannot explain their approach to addressing this implicit bias, ask more questions at a minimum. This should also a line item on the vendor selection criteria list.

5.6   AI REFERENCE ARCHITECTURES

"Operations keep the lights on, strategy provides a light at the end of the tunnel, but project management is the train engine that moves the organization forward." –Joy Gumz.[15]

> The reference architecture also embodies accepted industry best practices, noting numerous criteria specific to the industry of the product being developed, as well as often listing required functions, recommended processes to follow, and suggesting the optimal delivery method for specific technologies.[16]

Reference architectures embody accepted IT industry Best Practices and guidelines for the implementation of complex technology solution. They are designed to assist all parties involved to collaborate and effectively communicate during an IT implementation project. Frequently asked questions are identified and documented, resulting lower project risks with fewer delays and errors by learning from the knowledge of previous successful projects.[17]

There are a number of AI reference architectures and new ones continue to be developed. As with certain highly technical areas discussed, it is important that management be aware of the structure and make sure those tasks with the development and delivery of the AI suite of products and solutions understand and use the best practices associated therein.

This is likely a CAIO and his or her team can develop. Management need not know details but assure it is part of governance and policy. These models, like others, need to be *future-proofed*.

NOTES

1. "Quality Is Never an Accident." AZ Quotes. https://www.azquotes.com/quote/254792? ref=quality#google_vignette. Accessed October 8, 2025.
2. "Dot Bomb 2.0—ai Style." The Rapid Response Institute. https://therrinstitute.com/ dot-bomb-2-0-ai-style/. Accessed May 1, 2025.
3. "How Many AI Companies Are There in the World? [2025]." Ascendix. https:// ascendixtech.com/how-many-ai-companies-are-there/. Accessed May 1, 2025.
4. "Dotcom Bubble Definition." Investopedia. https://www.investopedia.com/terms/d/ dotcom-bubble.asp. Accessed May 1, 2025.
5. "What Is an Affinity Diagram?" ASQ. https://asq.org/quality-resources/affinity. Accessed May 1, 2025.
6. "When AI Thinks It Will Lose, It Sometimes Cheats, Study Finds." Time. https:// time.com/7259395/ai-chess-cheating-palisade-research/?utm_source=newsletter. genai.works&utm_medium=newsletter&utm_campaign=deepseek-goes-open-source-ai-is-cheating-and-humanoid-robots-are-hitting-the-market&_bhlid= 47e7ffdd6c7b2575f8fb42057e748dd1f6bab529. Accessed July 27, 2025.
7. "Key Skills to Look for in an AI Developer." Information Week. https://www. informationweek.com/machine-learning-ai/key-skills-to-look-for-in-an-ai-developer. Accessed May 28, 2025.
8. "Navigating the Data Minefields: Management's Guide to Better Decision-Making." CRC Press. https://www.routledge.com/Navigating-the-Data-Minefields-Managements-Guide-to-Better-Decision-Making/Shemwell/p/book/9781032677934#top. Accessed May 28, 2025.
9. "ISO/IEC 42001 Artificial Intelligence Management System." Standards Explained. https://standardsexplained.com/iso-iec-42001-artificial-intelligence-management-system/. Accessed May 28, 2025.
10. "Data Bias: The Latent or Unobserved." The Rapid Response Institute. https:// therrinstitute.com/data-bias-the-latent-or-unobserved/. Accessed May 28, 2025.
11. "Why Corporate Initiatives Fail." The Rapid Response Institute. https://therrinstitute. com/why-corporate-initiatives-fail/. Accessed May 28, 2025.
12. "sycophancy." Merriam-Webster. https://www.merriam-webster.com/dictionary/ sycophancy. Accessed
13. "Sycophancy in GPT-4o: What Happened and What We're Doing about It." OpenAI. https://openai.com/Index/sycophancy-in-gpt-4o/?utm_source=newsletter.genai.works& utm_medium=newsletter&utm_campaign=secret-ai-experiment-manipulates-reddit-users&_bhlid=38bf775f68087a179888e881d63d27d5a0763210. Accessed April 30, 2025.
14. "OpenAI Model Spec." OpenAI. https://model-spec.openai.com/2025-04-11.html. Accessed April 30, 2025.
15. "153 Project Management Quotes to Inspire You in 2025." dpm. https:// thedigitalprojectmanager.com/project-management/project-management-quotes-inspiration/. Accessed October 8, 2025.
16. "What Is a Reference Architecture?" BAE Systems. https://www.baesystems.com/ en-us/definition/what-is-a-reference-architecture. Accessed October 8, 2025.
17. "What Is a Reference Architecture?" Hewlett Packard Enterprise. https://www.hpe. com/us/en/what-is/reference-architecture.html. Accessed April 30, 2025.

6 Capstone—Detailed AI Models Under Development

The way to get started is to quit talking and begin doing.

—Walt Disney[1]

This is the capstone chapter. In addition to the Market Assessment AI Initiative described in some detail (Section 4.3.2.1 'Market Assessment' of Chapter 4), we delineate six additional actual model cases, four of which are the solutions of the author's organization and its partners. These include auditing, two examples of selection predictions, commodity future price influences, process manufacturing, cross-cultural team training, and organizational safety assessment. All of this author's solutions only use public data that is readily available and not subject to other copyright materials, such as website. Our data sources are public filings, government data, etc. and while we do not dispute the ownership of this data (census) is routinely used by third-party analysts and others.

The intent is to pull together the sum total of this book and provide best practices as well as a better understanding of the AI development processes that are currently available to organizations of all sizes and amounts content knowledge. Hopefully, these works in progress will give the readers ideas that may be useful in their organizations.

6.1 GENERIC ROADMAP

As part of our development of Chapter 3, Transformational Technologies: *Adoption and Integration in Operating Plants* of our 2023 book, *Smart Manufacturing: Integrating Transformational Technologies for Competitiveness and Sustainability (1st ed.)* we used an affinity diagram (Section 5.2.1.2 'Populating the Wheel' of Chapter 5) to capture thoughts about six categories. This 'working' data was not published at that time and it has been updated to include the use of AI to transform manufacturing. This roadmap framework is applicable to most other sectors and firms as well.

- **Strategic Options**—Focused on AI selection and implementation approaches, management must develop a deep and substantial understanding about how AI will benefit the organization and its Risk Profile.
- **Operations**—Identify how will AI be used in all aspects of revenue generation and cost management processes.

DOI: 10.1201/9781003646914-6

- **Competitiveness**—Determine the value of AI to better position the firm in the market, especially with respect to its current and possible future competitive landscape.
- **Sustainability**—Organizational sustainability as well as technology refresh or future-proofing.
- **Intangibles**—There are more latent variables at work then most executives understand. We have shown how AI can be used to better understand and management aspects such safety, productivity, and so on (Section 1.12.2 'The Latent Construct' of Chapter 1).
- **Performance Forecasting**—Most critical decisions are about the future and realizing expected results. This is an inherently difficult task and we posited several approaches in this book, Future Commodity Price Influences (Section 6.4 'Case Three—Oil & Gas Sector Future Price Influences' of Chapter 6) and Market Assessment (Section 4.3.2.1) among others.

As shown Figure 6.1, this body of organizational knowledge, history, and desire future is built upon the appropriate AI technology foundation. The cornerstone of the model, the Risk Mitigation Strategy is perhaps the most important building block of modern institutions.

We have found this process effective in workshop environments, where knowledgeable contributors can flush out thinking about these and other components of the model. Subsequently, this line of thought can be captured strategically as well as used in daily operations.

It can help to compare your organization's roadmap with peers and even those in other sectors. Generally, there is enough published information as well as knowledge of the competitive landscape. AI tools can also be used for this straightforward review.

6.1.1 OPERATIONAL EXCELLENCE

We have written extensively on this subject for many years, including in our recent book, Navigating the Data Minefields: Management's Guide to Decision-Making. Therefore, this section is a brief high-level frame of the model that plays a major role in all operations and their decision support systems.

Once we understand our roadmap as shown above, we need to operationalize it into daily business processes. Most businesses use an Operations Management System (OMS) as their actionable model. In conjunction with several firms in the energy sector, we researched industry efforts and developed the following model.

Our Smart OpEx OMS solution has eight action items as a function of four measurable processes.[2] These are depicted in Figure 6.2 with focal points checked as appropriate.

Most of the eight action items have been discussed already. However, there are two that require additional information to put their importance into perspective.

- **Privilege to Operate**—Most businesses require some degree of regulation. Those in the critical infrastructure sector are highly regulated due to the

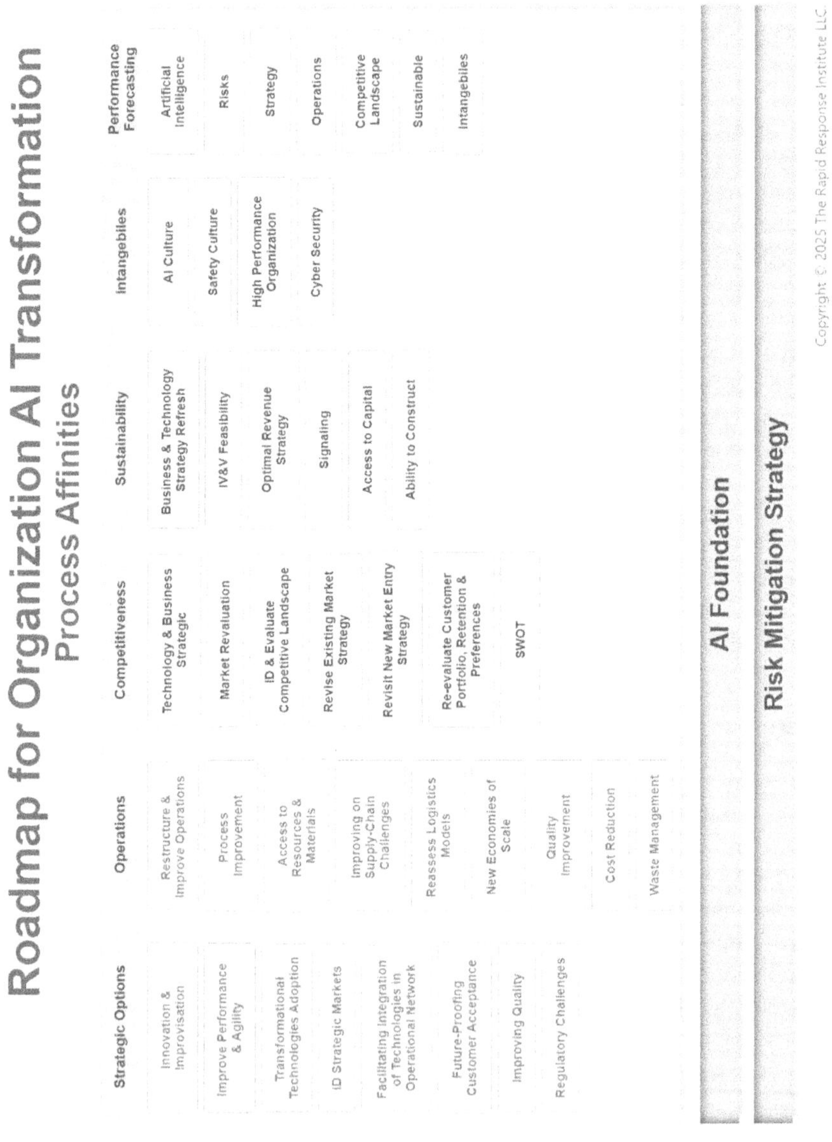

Roadmap for Organization AI Transformation
Process Affinities

Strategic Options	Operations	Competitiveness	Sustainability	Intangibles	Performance Forecasting
Innovation & Improvisation	Restructure & Improve Operations	Technology & Business Strategic	Business & Technology Strategy Refresh	AI Culture	Artificial Intelligence
Improve Performance & Agility	Process Improvement	Market Revaluation	IV&V Feasibility	Safety Culture	Risks
Transformational Technologies Adoption	Access to Resources & Materials	ID & Evaluate Competitive Landscape	Optimal Revenue Strategy	High Performance Organization	Strategy
ID Strategic Markets	Improving on Supply-Chain Challenges	Revise Existing Market Strategy	Signaling	Cyber Security	Operations
Facilitating Integration of Technologies in Operational Network	Reassess Logistics Models	Revisit New Market Entry Strategy	Access to Capital		Competitive Landscape
Future-Proofing Customer Acceptance	New Economies of Scale	Re-evaluate Customer Portfolio, Retention & Preferences	Ability to Construct		Sustainable
Improving Quality	Quality Improvement	SWOT			Intangibles
Regulatory Challenges	Cost Reduction				
	Waste Management				

AI Foundation

Risk Mitigation Strategy

FIGURE 6.1 AI transformation roadmap structure.

-- AI Enabled --
Smart OpEx Operations Management System (OMS) Framework

	People	Process	Plant (Facility)	Performance
Leadership	✓			
Organization	✓			
Risk		✓		
Processes & Procedures		✓		
Assets			✓	
Optimization			✓	
Privilege to Operate				✓
Results				✓

✓ Primary Focus

FIGURE 6.2   AI-enabled OMS framework.

nature of the enterprise and the potential downside from failures. Often the subject of routine and spot checks or audits, the right to remain in business can be withheld. We see this when aircraft are grounded following a major incident. Clearly, this falls under the umbrella of Performance.

- **Results**—Likewise, organizations are responsible to shareholder for financial performance. They are also held to high standards by other stakeholders including local communities, clients, employees, and supply chain participants.

AI can play a vital role enabling firms to not only address these issues but seen as a high performance by equity markets as well as a great corporate citizen by others. Examples include makers of smart devices, medicine, and transportation providers. These companies can fall from grace and airline companies often do as well as others perceived to be environmentally unfriendly. AI has the potential to raise all and allow some to shine with super high performance and a deserved reputation.

AI-Enabled Operational Excellence Transformation

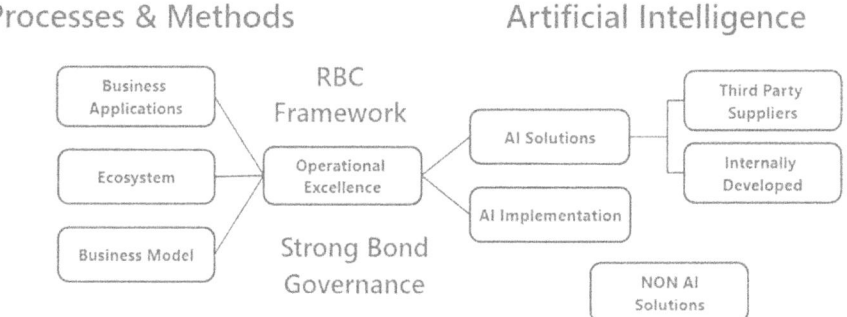

FIGURE 6.3   AI-enabled operational excellence cultural transformation.

In Section 6.5 'Case Four–Smart Manufacturing: The Re-Refining Process' of Chapter 6, we review an actual project current underway during the writing of this manuscript. It goes into detail about these processes and how AI is used to achieve additional value.

One of the processes we are undertaking if not simply the automation or digitalization of the process industry firm, but to use AI to transform the way it does business. As shown in Figure 6.3, Operational Excellence is achieved build on a strong governance model and our Relationships, Behaviors, Conditions model of human interaction.

In earlier models, we identified two major components, Processes & Methods, and Enabling Tools such as AI. AI will drive not only internal business (process) applications but those of its ecosystem or supply chain and client base. Enabled by a Body of AI Knowledge (developed throughout this book) drawn from all sources and effectively implemented, organizational transformation into one of excellence using AI is assured.

Finally, keep in mind that there will be some software applications that many not be appropriate to convert or may later in the enterprise-wide AI transition. These applications may or may not interact with AI.

Sectors and individual organizations have constantly undergone, sometimes radical transformations in a short period of time. Best practices are well understood and work well unless they are poorly or casually implemented, often treated as short-term initiatives.

As a general rule, change management models are appropriate for AI transitions. Strong, visible, and consistent leadership is the singular catalyst for success. The rest are all tools for leadership.

6.1.1.1 AI Body of Knowledge

It is critical that organizations internalize a body of knowledge as a major component of its Core Competency. While third-party content and technology expertise is essential to remain current, this requirement cannot be outsourced!

It is not necessary to establish an AI Office; however, it is important that AI knowledge be codified into the organization. It will be the responsibility of CAIO to put processes and procedures in place enabling a virtual approach.

As shown in Figure 6.4, this must be aligned with the overall strategy as enabled by the OMS and culture. Appropriate expertise must be developed internally and supported by outside expertise, as necessary.

The organization must develop a core understanding of AI, just as it does about the nature of the mission. Moreover, management is responsible for assure employees and other members of the supply chain are trained and remain current in technologies and the processes they enable. This includes a world class best practice implementation of AI solutions.

Finally, the AI maturity of the firm as well as AI technologies must be taken into consideration and shortcomings candidly addressed. It is critical that all speak the same language, in other words a common vocabulary consistent with generally accepted terminology.

Our strong recommendation is that the organization through the CAIO formally structure AI, so it is clear to all that management is committed to the AI direction. Most initiatives fail because they lack this formalized model. If AI is perceived as this important, as they saying goes 'failure is not an option,' so act accordingly.

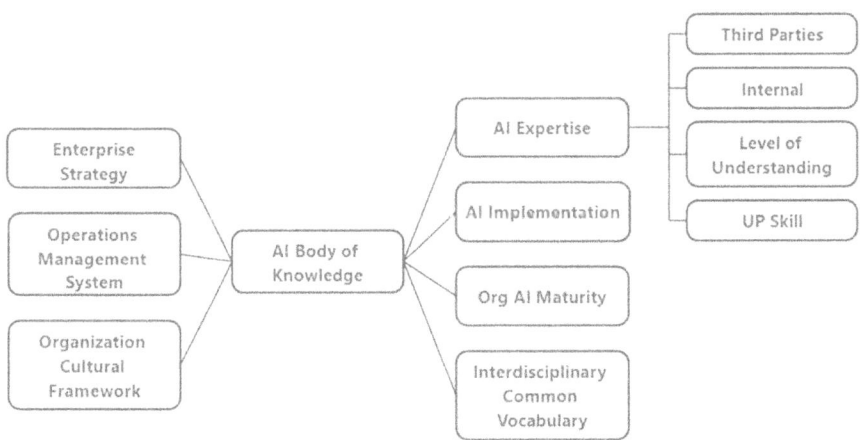

FIGURE 6.4 AI virtual AI center of excellence model.

6.1.1.2   AI Service Level Agreement

Most are familiar with IT Service Level Agreements (SLAs). These are common with technology contracts and are designed to assure an agreed upon level of performance from systems used to manage critical aspect of the enterprise. We believe that AI deployments require SLAs; however, there are some unique AI aspects of these legal agreements. A few things specific to AI issues to consider include:

- **Service Description and Scope**—Develop specific and detail list of AI capabilities provides as well as the intended use, supported platforms, and limitations.
- **AI-Specific Performance Metrics**—In addition to typical SLA requirement of uptime and response times, consider including:
  - **Model Accuracy**—We have discussed this issue in detail including biases. Expectation must be realistically set given the maturity of AI and their attainability measurable.
  - **Response Time**—AI complexity should be realistically understood and agreed upon.
  - **Availability**—Traditional high-level availability may not be possible given AI constraints.
- **Data Ownership and Privacy**—This can be a tough problem, and likely ownership will be consistent with typical SLAs. However, the use of data for AI training needs to be understood, agreed to, and the ownership of the output established in advance.
- **Support and Maintenance**—Specific to the AI model, in addition to an agreed upon maintenance program, model update processes and frequency must be understood as well as the process for addressing model issues in a timely fashion.
- **Liability and Indemnification**—The output of AI models can put the organization at risk if there are errors, omissions, or other issues. This is an area that must be well throughout as a function of the level of risk being taken.

As with all agreement of this nature, it must be aligned with the governance models and operational processes and procedures. Annual updates are advised as well as a review when the AI environment changes in any significant manner.[3]

The opinions offered herein are the thinking of a non-lawyer writer. Organizations should consult with appropriate legal counsel when drafting this and other legally binding agreements when implementing artificial intelligence.

6.2   CASE ONE—AUDITING USING AI

In late 2024 Elon Musk and his team assessed the proposed federal government 1,547-page budget. This shed some interesting light on the management and understanding of very large documents. Surprising to many, he quickly developed an immediate grasp of the document and its contents.

Perhaps the process of throwing massive documents at the feet of decision-makers at the very last minute has ground to an ignominious end. While I never saw a definitive statement, it appeared that Musk used AI and defeated a process designed to hide spending from the electorate. Enter a new era of the greatest disinfectant (sunlight) and accountability?

Unless otherwise noted, the following is largely taken from our February 10, 2025, blog, *Process Audit-Musk Style: A Systems Analysis Solution.*[4] This information is based on public data regarding DOGE assessments at the US Treasury department. We believe this is a high value use of AI, and project this body of knowledge into the general auditing processes well known to most organizations.

According to the American Society for Quality (ASQ), auditing has three components[5]:

- **Process Audit**—Verification that processes are working properly within established limits.
- **Product Audit**—Assess whether or not products and services conform to accepted standards and requirements.
- **System Audit**—Assessment of management systems, i.e., OMS (Section 3.2.1 'Operations Management System' of Chapter 3).

The principal objective of the systems analysis phase is the specification of what the system needs to do to meet the requirements of end users. Isn't what Musk has done the holistic integration of data acquisition and initial assessment of the three components of an Audit?

Key take aways, this pundit took include that crawling through decades of data:

- Three subsystems were mapped quickly as were payment flows across agencies. Patterns were revealed that even career individuals did not know existed.
- $17 billion in redundant programs were detected and are growing.
- 'The beautiful thing about payment systems,' noted a transition official watching their screens, 'is that they don't lie. You can spin policy all day long, but money leaves a trail.'
- That trail led to staggering discoveries. Programs marked as independent revealed coordinated funding streams. Grants labeled as humanitarian aid showed curious detours through complex networks. Black budgets once shrouded in secrecy began to unravel under algorithmic scrutiny.
- By 6 AM, Treasury's career officials began arriving for work. They found systems they thought impenetrable already mapped. Networks they believed hidden already exposed. Power structures built over decades revealed in hours.
- Their traditional defenses—slow-walking decisions, leaking damaging stories, stonewalling requests—proved useless against an opponent moving faster than their systems could react. By the time they drafted their first memo objecting to this breach, three more systems had already been mapped.[6]

This is only a partial list of value that can be derived using AI to audit a large organization with perhaps millions of digital papers and online databases. This is a game changer in the way this process has historically been very manual (even after the advent of high-performance computing). Auditors and subject matter experts must review diverse sets of materials, find gaps, and make decisions based on incomplete, missing and possibly incorrect/corrupt data.

This can be a Creative Destruction moment in this professional services sector. It supports comments made that junior white-collar workers who are at the most risk from AI.

At a recent meeting, my client asked me how many programmers we used for projects. My response was never more than three, but we augmented specific skill needs, as necessary. The days of armies of programmers and consultants are ending. Tools supporting core internal capabilities return IT core competency to organizations. One can expect to see outsourcing fall and perhaps rapidly in the near future.

Augmenting key skill sets and capitalizing on true knowledge for projects have been the historic role of professional services. Those organizations who can return to their roots can do well albeit with significantly different business models than those of today.

Not everyone will be happy with this model. One surmises that in Washington, DC, this approach will expose pork and even nefarious behaviors. The same may be true in private industry and the impact of large auditing and consulting engagements may be negative. It is, however, providing sunlight into these types of processes, disinfecting the negative aspects, and adding direct value to organizations.

6.3 CASE TWO—AND THE WINNER IS!

Humans love a winner and betting on a winner is as old as recorded history. Not surprisingly, AI has become a tool for predicting the future. One algorithm taxonomy group is even labeled, Predictive AI. We have long used using tools to guess or predict the future. Statistical predictive analytics is a forerunner of Predictive AI.

Two contemporary elections offer some lessons learned when using Predictive AI. Two different outcomes, yet similar lessons regarding information and data bias.

6.3.1 NFL DRAFT, 2025

One month before the 2025 National Football League draft, USA TODAY used Microsoft Copilot AI Chatbot to predict the first round (Table 6.1). The article caveat,

> Copilot wasn't quite up to the task of creating an entire first-round mock by itself. It was able to mock the top 10 selections. Still, after that, it kept struggling to stick to the draft order while also including several players from the 2024 NFL draft in its projections.[7]

On April 24, 2025, the actual draft was held. The results follow.[8]

In some ways, this was not a big challenge for AI, since the pundit ratings of these players have been discussion points for a long time, and hence the data AI had to learn from reflected those biases. According to Forbes, Sanders has an attitude problem that was well known and documented. One wonders if this factored into the AI training, was it simply overlooked or a product of AI Hallucination?[9]

Perhaps, this variable was not weighted properly. Out of the ten forecast selections, two were in error (negatively) by 50% and one by in excess of 5,000%. We did

TABLE 6.1

2025 NFL Draft: AI Selection vs Actual Draft (Developed by The Rapid Response Institute

Team	Mock Round One (AI Picks, March 2025)	Actual Draft (April 24, 2025)	Differential
1. **Tennessee Titans**	Cam Ward, QB, Miami (FL)	Cam Ward	0
2. **Cleveland Browns**	Abdul Carter, EDGE, Penn State	Travis Hunter	+2
2a.		Abdul Carter	−1
3. **New York Giants**	Shedeur Sanders, QB, Colorado (actual 144th pick)	Abdul Carter	−141
4. **New England Patriots**	Travis Hunter, WR/CB, Colorado	Will Campbell (no data here as the model is the function of Travis Hunter)	−
5. **Jacksonville Jaguars**	Mason Graham, DT, Michigan	Mason Graham	0
6. **Las Vegas Raiders**	Ashton Jeanty, RB, Boise State	Ashton Jeanty	0
7. **New York Jets**	Armand Membou, OT, Missouri	Armand Membou	0
8. **Carolina Panthers**	Tetairoa McMillan, WR, Arizona	Tetairoa McMillan	0
9. **New Orleans Saints**	Tyler Warren, TE, Penn State (actual 14th pick)	Kelvin Banks Jr.	−5
10. **Chicago Bears**	Jalon Walker, LB/EDGE, Georgia (actual 15th pick)	Colston Loveland	−5

not calculate all selections, but that might be an interesting exercise for the NFL or other analyst, learning for the AI selection algorithm.

Question, if your decision requires the selection of ten answers to rank ordering, is the AI response acceptable? In this case the solution was easily validated; however, this is not always the case.

6.3.2 THE ELECTION OF POPE LEO XIV

On April 21, 2025, Pope Francis passed away, triggering the search for a successor. This centuries old selection process is conducted in secrecy within a limited time frame by only a small number of voters (Catholic Church Cardinals). A small sample size for any statistical calculation.

- According to one researcher using AI, Michele Re Fiorentin, a physicist at the Polytechnic University of Turin, and University of Madrid, one of the modelers, "the model probably missed Prevost as a likely pope because it didn't consider political and geographical factors that played a role in the election. Lacking that information, he says, 'is a major shortcoming of our model.'"[10]

Statisticians have developed approaches for Small Numbers (small data sets), that are applicable for soft variables such as discussed herein. The details are beyond the scope of this book, but interested readers can review the cited material.[11] AI developers are advised to familiarize themselves with these statistics.

In both the Vatican and NFL cases, society-based data played a bigger role than the political and functional skill sets of contenders. As shown in the model developed to counter the poor performance of political polling of predictive shortcoming of the US presidential election held November 5, 2024 (Section 4.3.2.1.1 'US Presidential Election Surveys' of Chapter 4), older, and generally accepted behavioral models may need an overhaul in the AI era.

6.4   CASE THREE—OIL AND GAS SECTOR FUTURE PRICE INFLUENCES

We all know that energy underpins the global economy. Therefore, the energy price and expected price in the future play a major role in decision-making at all levels, from government policy to individual vacation plans.

For this model, we use the price of oil. However, we believe the basic model applies to all commodity types, including non-energy, i.e., food, precious metals, etc.

6.4.1   DECISION CYCLE-TIME

In our 2009 White Paper, *Rapid Response Management: Thriving in the New World Order*, we coined the term, *Velocity of Information* (VI) which we defined as "similar to the economic theory, Velocity of Money, it is the frequency at which information is exchanged." Initially focused on the *Sense-and-Respond* inventory mechanism system at the time driven by the new RFID tagging solution. We went on to liken it a function of the convergence of exponentials, to

> Moore's Law stating that the number of transistors on a microprocessor doubles every 18 months; Metcalfe's Law stating the usefulness of a network equals the square of the number of users; and Gilder's Law stating bandwidth rises three times faster than the power of the computer; unleashes untapped energy that we are just beginning to understand.[12]

VI is not the time it takes commodity pricing and their futures to be reflected by markets. These are posted publicly in real time. Rather VI is a function the time it takes an organization to internalize and propagate new information for the appropriate needs of the firm. This can vary by region, organization department, and even at the individual level.

This process correlates directly with organizational decision cycle times. Those firms with a *slow-twitch* information muscle are at a disadvantage to those with a *fast-twitch*. This can be seen as a metric of "organizational metabolism."

AI is enabling this acceleration exponentially new velocities, and we expect will impact on decisions made for equity, bond, and commodity markets among financial, operational, safety, and productivity measures. Figure 6.5 is a depiction of AI-enabled increase in the velocity of information. Typical OT-IT systems are being replaced with tightly integrated 'OTIT' systems as described in Section 3.2.1

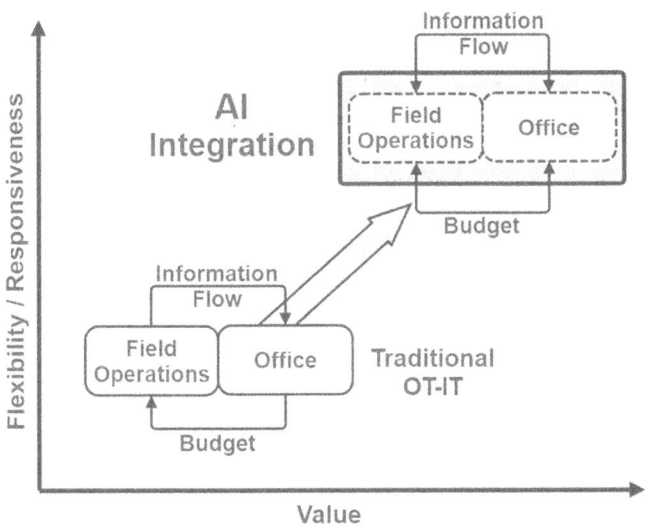

FIGURE 6.5   AI integration: Enterprise decision-making process.

'Operations Management System' of Chapter 3. Subsequently decisions are made faster and in some cases such as an ongoing field operational issue such as a fire or explosion, can result in saving lives as well as less property damage and downtime (Section 1.8.2 'Complex Adaptive Systems' of Chapter 1).

We continue to redefine 'timely information.' We used to talk in terms of Internet Dog Years, Near Realtime, and process cycle time in manufacturing output terms. Now the demand for high performance can be measured at our speed of thought. Expect this decision-making frequency period to continue to increase.

6.4.2   FUTURE CRUDE OIL COMMODITY PRICING

This is a preliminary model for teaching discussion purposes only. Neither the author nor the publisher takes any responsibility for its accuracy or appropriateness. It is meant as information and decisions should not be taken based on outputs from this prototype.

The price of a barrel of crude oil (benchmark West Texas Intermediate aka WTI) affects everyone on the planet. Whether filling the gas tank of our automobile, purchasing heating oil for our home or traveling on a commercial airline, this metric is highly visible and its impact on individuals is immediate. WTI is a US metric and there are other similar ones used in various parts of the world. This fundamental model remains the same regardless the crude oil classification used, i.e., Brent, Dubai, Bitumen.

Businesses, such as trucking companies and airlines consume massive amounts of fuel and other products manufactured from crude oil. The value of their business and its cash flow often depend on successful hedging of this commodity in the marketplace.

Predictions for this important global commodity future price point and trends are the focus of almost every major financial service and banking provider's research department. The oil and gas sector economists have the most to win or lose with their predictions (both short and long term) as capital investments can take years or more to reach break-even and longer before profitability. CAPEX are G0-No Go based on future price expectation. Finally, a bank's willingness to loan is based on this metric and the value of an oil and gas firm is based on its proven reserves ("those quantities of petroleum anticipated to be commercially recoverable by application of development projects to known accumulations from a given date forward under defined conditions") at a predicted price point.[13] Talk about risk!

Not quite a 'black art,' but being profitable in the Black Oil aka Texas Tea business depends on beating the commodity roulette wheel. Long a challenge, this task is even more challenging today's business and geopolitical environment.

Regarding price volatility, "It's not A plus B equals C anymore. There are like nine equations here. There are so many things going on at once that you pull on a string, you don't know where the other end of the string is going to be," said Dan Pickering, chief investment officer at Pickering Energy Partners.[14] This appears to be a set of Difference Equations and a likely candidate for artificial intelligence assessment.

We expanded the basic EIA Crude Oil Supply and Demand pricing model to include intangible or soft variables typically not incorporated into future price prognostication models.[15] Figure 6.6 depicts this expanded process and subsequent definition of the data required as well as the set of function equations that frame a *set of simultaneous equations* (Appendix I—Glossary of Terms) in the next phase of the project follow.

The model must use a number of variables, including latent variables as depicted below. This is a complex problem and from this writer's perspective the fidelity of the model cannot be compromised. That said, humans struggle with complexity, and this approach seeks to address this conundrum.

6.4.2.1 Model Functions

The following functions are the core of this influence model. Mathematical functions are defined as, "an expression, rule, or law that defines a relationship between one variable (the independent variable) and another variable (the dependent variable). Functions are ubiquitous in mathematics and are essential for formulating physical relationships in the sciences."[16] They are not equations, nor do they have enough detail to enable calculations. Finally, these are provided as a First Draft of this work, and the final product may be somewhat different.

These nine functions embody this comprehensive model and are composed of almost 100 sub variables including almost 40 latent variables. In the mind of this developer, this is the perfect model for the use artificial intelligence.

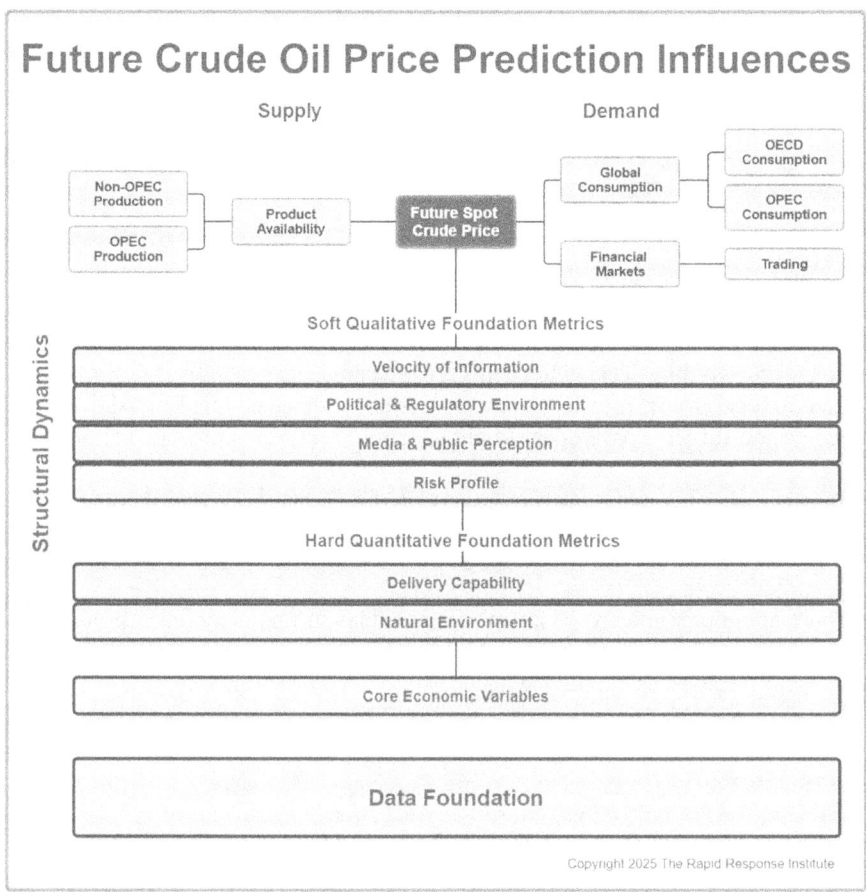

FIGURE 6.6 Future crude oil price prediction influences.

Following this set of functions, more detailed definitions of components are provided. They form the basis for set of simultaneous equations to be developed in the near future.

$$CEV_{\bar{x}} = f\left(SP_C + SP_T + WTI_{PS} + WTI_{APS} + CDP_\delta + HASP_{\bar{x}} + GAS_P + REF_{AC}\right)$$

$$FIN_{\bar{x}} = f\left(DOI_A + RPC_{30} + FP_U + FP_{MM} + COR_{CC} + CI_{DOW} + COM_{DOW}\right. \\ \left. + COR_{CF} + HRA_{NE} + PFP\right)$$

$$BAL = f\left(\left(FS_{NO} + f\left(WTI_{APS}\right)\right)\right)$$

$$PCA = f\left(P_T + FS_T + HPA_{\bar{x}} + |PCA_C|\right)$$

$$CON_G = f\left(CON_{NO} + CON_O\right)$$

$$\mathrm{DEL} = f\left(\mathrm{DIS_H} + \mathrm{LAB_S} + \mathrm{TER} + \mathrm{CYBR} + \mathrm{UNP} + \mathrm{EP} + \mathrm{EA}\right)$$

$$\mathrm{MPP} = f\left(\mathrm{MED_{\bar{x}}} + \mathrm{PER_{\bar{x}}}\right)$$

$$\mathrm{EV_{NI}} = f\left(\mathrm{VI} * \mathrm{U_M}\right)$$

$$\mathrm{PRE_{\bar{x}}} = f\left(\mathrm{POL_{\bar{x}}} + \mathrm{REG_{\bar{x}}}\right)$$

Equation 6.1—Set of Future Crude Oil Price Influence Functions

Table 6.2 lists the description of each component and symbol for each of the Independent Variables. The components are labeled using generally accepted abbreviations and/or terminology. If the reader does not understand this language, a quick search should suffice. Space does not allow detailed definition for each variable.

> The Latin letter f is used in mathematics to represent the name of a generic math function. Typically, the symbol is used in an expression like this: $f(x)$. In plain language, this expression represents the function named f that takes in the variable x as input.[17]

There are approximately 30 latent *soft* variables in this model. As discussed in Section 1.12.2 'The Latent Construct' of Chapter 1, we can quantitatively manage these usually viewed as qualitative data sets.

6.4.2.1.1   Linear Function
According to Forbes,

> Hard metrics are the data and charts you see. Soft metrics are intangible and hard to measure because of causation and correlation. An example of a soft metric would be having a customer on social media say positive things about your company. This may lead to an increase in sales, but it's unclear if it was specifically from causation or correlation.[18]

For this influence model, we use simple Linear Functions,

> These have the form (y = mx + b) where (m) and (b) are constants representing the slope and y-intercept, respectively. Here, the relationship between x and y is direct and steady, symbolizing uniform growth or decay. This concept is at the heart of many mathematical equations and can be represented as $= f(X)$. The function notation $f(X)$ doesn't imply multiplication, but rather it signifies that y is the output of the function for the input X.[19]

We have argued for over 20 years that intangible (soft) variables are real and can be incorporated into financial and risks assessments. Our Economic Value Proposition Matrix® (EVPM) has been used for many assessments since its development beginning in 2004, including measuring the value of security in geopolitically unstable areas.[20] As part of this AI development effort, we are expanding the use of that technology prompting for insight into the data of others with other projects. For example, data from published results of similar efforts will be rolled into future EVPM models. In

TABLE 6.2
Model Function Components

Core Economic Variables—(CEV$_{\bar{x}}$)

Component	Symbol	Component	Symbol
Closing Spot Price	SP$_C$	Changes in GDP	GDP$_\delta$
Spot Price Trend	SP$_T$	Historical Spot Price Forecast Accuracy	HASP$_{\bar{x}}$
Closing WTI Future Price Spread	WTI$_{PS}$	US Retail Gasoline Price	GAS$_P$
WTI Price Spread (Ave 6 month)	WTI$_{APS}$	Refiner Acquisition Cost of Crude	REF$_{AC}$

Financial Markets—(FIN$_{\bar{x}}$)

Component	Symbol	Component	Symbol
Average Daily Open Interest (Futures)	DOI$_A$	Commodity index assets under management & Dow Jones UBS price index	CI$_{DOW}$
Ratio Puts/Calls	RPC$_{30}$	Composition of the Dow Jones UBS commodity index	COM$_{DOW}$
Future Positions Taken by Crude Users	FP$_U$	Correlations between daily returns on crude oil & financial investments	COR$_{CF}$
Future Positions Taken by Money Managers	FP$_{MM}$	Average Historical Reaction to Negative Events	HRA$_{NE}$
Correlations between daily prices changes of crude & other commodities	COR$_{CC}$	Perceived Future Price	PFP

Trading—(BAL)

Component	Symbol	Component	Symbol
Non-OPEC Forecast Supply	FS$_{NO}$	WTI Price Spread (Ave 6 month)	WTI$_{APS}$

Note: Including OPEC trading may yield a better result and is being explored.

Projected Crude (Supply) Availability—(PCA)

Component	Symbol	Component	Symbol
Non-OPEC (OECD) Production	P$_{NO}$	OPEC Forecast Supply	FS$_O$
OPEC Production	PO	Total Forecast Supply	FS$_T$
Total Production	P$_T$	Historical Sport Price Forecast Accuracy (Weighted Average)	HPA$_{\bar{x}}$
Total Production Capacity	PC$_T$	OPEC Forecast Supply	FS$_O$
Non-OPEC Forecast Supply	FS$_{NO}$	Historical PCA Compared with Historical Average (absolute value)	|PCA$_C$|

Global (Demand) Consumption—(CON$_G$)

Component	Symbol	Component	Symbol
OECD Inventory	IN$_{NO}$	OPEC Inventory	IN$_O$
OECD Crude Oil Consumption	COC$_{NO}$	OPEC Consumption	COC$_O$
Global Oil Consumption	COC$_W$	Global Oil Consumption	COC$_W$

(Continued)

TABLE 6.2 (*Continued*)
Model Function Components

		Delivery Capability—(DEL)	
Component	**Symbol**	**Component**	**Symbol**
Human Made Disruptions	DIS_H	Unplanned down time	UNP
Labor Unrest/Stoppage	LAB_S	Environmental Protests	EP
Terrorism	TER	Equipment Availability	EA
Cyber Attacks	CYBR		

		Media & Public Perception—(MPP)	
Component	**Symbol**	**Component**	**Symbol**
Media	$MED_{\bar{x}}$	Public Perception	$PER_{\bar{x}}$

		Expected Value of New Information—(EV_{NI})	
VI (Velocity of Information)	**Symbol**	**U_M (Utility/Satisfaction Max)**	**Symbol**
Contextual Information (Weighted Average)	$CI_{\bar{x}}$	Marginal Utility of a Unit of Information	U_1
Supply of New Information (Weighted Average)	S_{NI}	Marginal Utility of Two Units of Information	MU_2
		Marginal Utility of Three Units of Information	MU_3
		Marginal Utility of N Units of Information	MU_N

		Political & Regulatory Environment—($PRE_{\bar{x}}$)	
Political Component	**Symbol**	**Regulatory Component**	**Symbol**
Propensity to Support Oil Sector	P_{SOS}	Trends	REG_{TRE}
Geopolitical Stability	GS	Tax	REG_{TAX}
Legal Structure	LS	Environment	REG_{ENV}
Policy: Energy	POL_E	Transportation	REG_{TRAN}
Policy: Trade	POL_{TRD}	Energy	REG_E
Policy: Tax	POL_{TAX}		
Policy: Environment	POL_{ENV}		

the past, these risk-adjusted numbers were largely the input from internal organization workshops and thus by definition limited to the personal experiences of a small group. We believe this broader use will increase the validity and reliability of the model.

6.4.2.2 Planned Enhancements

This model is at the Proof of Concept (PoC) stage and there is a great deal of work yet to do. From my perspective, the problem definition and the structural equation modeling are the gating components.

To date, we have developed a very detailed model as well as the first group of Prompts. We are exploring the use of "Physics-Informed Neural Networks (PINNs), a type of artificial intelligence that incorporates physical laws into its learning process."[21]

6.5   CASE FOUR—SMART MANUFACTURING: THE RE-REFINING PROCESS

In our 2025 book, *Navigating the Data Minefields: Management's Guide to Better Decision-Making* we addressed the operational side of the smart manufacturing foundation, Operational Technologies—Information Technologies (OT-IT).

> Many readers may not be familiar with Operational Technology, which integrates, monitors, manages, and secures industrial processes and safety.
>
> A partial list of the types of applications they are found include Robots, Industrial Control Systems (ICS), Supervisory Control and Dat Acquisition (SCADA), Programable Logic Controllers (PLCs), Computer Numerical Control (CNC), Distributed Control Systems (DCS), Emergency Management Systems and Medical Equipment. Generally, they acquired data using sensors and sometime remote sensors. The amount of (often real time) data collected can be massive. Its analysis, as you will see in this book, is likewise huge and can have an extremely high value proposition. On the other hand, IT is the more traditional use of computer processing such as finance, human resources, and other applications most are familiar with. Increasingly, real value comes from the integration such as found in manufacturing and inventory management.[22]

The emerging era of agentic AI will most likely change the way humans interact and collaborate in the Smart Manufacturing environment. Traditionally, AI assistants are rule based with limited independent decision-making capabilities. Agentic AI, on the other hand possesses significant autonomy. Achieving such decision-making autonomy requires a highly complex of different machine learning, natural language processing, and automation technologies working in a new way. The focus is on decision-making processes rather than content creation that are set to optimize a set of goals (Pareto Efficiency). Agentic AI then works independently. Moreover,

> For agentic AI to succeed, the models must have SMART (*S*pecific, *M*easurable, *A*chievable, *R*elevant, *T*ime-bound) [emphasis added] goals and sub-goals and know how to measure them. They must have the right contextual information—why are these goals important to the company, how do they drive revenues, etc. Finally, as managers, we need to establish feedback loops to adjust the models as we learn more about their performance.[23]

There is a lot of potential to use agentic AI in a large number of applications across a wide variety of public and private sectors. To be successful, these initiatives require a diverse human team of subject matter experts.

These teams will face all of the issues and problems cross-functional teams traditionally see, such as conflict, resource management as well as diverse personalities. Team building processes such as those found in cross-cultural training must be utilized if the maximum output from agentic AI is to be realized.[24]

Manufacturers will probably phase our lower skilled functions such as data collection and analysis, but higher-level decision supporting analysis will still be required. However, at some point agentic AI may make inroads into those needs as well. Perhaps this will mitigate the pending retirement of long-time experts.[25]

As part of its environmental strategy, the fossil fuel industry recycles used petroleum products. Typically referred to as oil Re-Refining, this process has a positive effect on the environment and oil can be re-refined many times.

This project is relatively new, and we are still building the Smart version as well as collecting internal client data. We expect, as above, a large number of variables.

Our expectation that processes such as Preventative Maintenance and facilities optimization are areas where AI can add immediate and sustained value. Effectively, oil becomes a renewable energy source.

As shown in Figure 6.7, these type organizations consist of two major components, the Core Foundation and Operational Processes. A brief definition follows.

Core Foundation

Policy—Organizational Governance requirements

Standards—Industry Standard such as API and non-Industry such as ISO management systems and other standards

Competency—Knowledge, Skills, and Abilities (KSA) of the combined workforce as well as its state of Training

Tools—Whatever is required to perform job functions including hand tools, software, and sensors, etc.

Operations

Process—High-Level Business Process that maybe across functions, departments, and even firms

Procedure—Subset of the overall process that is a series of steps to accomplish good practices

Relationship—Third parties perform most of the work in the field. There is a relationship between the operator and contractor as well between contractor and subcontractors

Task—is the Job to be performed at the worksite such as repair a pump, receive/ship product or Safety and Environmental Management System (SEMS) Audit

Sub Task—those major components of a Task

Figure 6.7 reflects all of these components and their relationships. Each generates significant data, even for a small facility.

We expect to initially use AI for predictive maintenance. According to Oracle, we expect our customer can gain the following value proposition in line with the experience of others[26]:

- **Reduces Costs**—"AI algorithms can also closely track a machine's energy consumption, detecting inefficiencies and suggesting steps to save money. They can even help trim labor costs by prioritizing maintenance work, thus reducing unneeded inspections, repairs, and replacements."
- **Limit Disruptions**—Faster and more accurate predictions than previous methods.

FIGURE 6.7 Generic AI smart OpEx-enabled process sector solution.

- **Increases Production**—According to a 2022 Deloitte Study, using predictive maintenance AI tools can boost productivity from 5% to 20%.
- **Improves Safety**—Preemptively making repairs can avoid putting employees at risk if the equipment becomes damaged as the result of failure.
- **Extends Equipment Lifecycle**—"AI-based data analytics tools can help prolong a mechanical asset's lifespan, boosting manufacturer uptime, productivity, and, ultimately, revenue."
- **Improves Quality Control**—"AI tools help improve product quality and consistency, minimizing defect rates and reducing production costs. When AI algorithms are trained on massive amounts of product specification data, they can find cracks in products, misalignments, and other issues."

We are following the roadmap previously discussed and believe for this organization, predictive maintenance is the best first step and will demonstrate to field personnel and the corner office that AI adds significant value. Once this first step is complete and fully operational, we will address other areas where AI/ML etc., can add value. This approach is consistent with the company's policies and maturity with sophisticated systems.

One of the takeaway items from this project is that even small operators can gain demonstrable value from AI and its derivatives. No process is too small given AI price points with the likelihood they will continue to decline as the technology matures and becomes mainstream.

6.6 CASE FIVE—AI-DRIVEN CROSS-CULTURAL SERIOUS GAMING

Perhaps the most exciting project we have underway is the transformation from our flagship serious cross-cultural teaming game to one where human teams no longer have to engage other human teams, but one where AI is the other collaborator. This increases flexibility and enables us to provide greater access to clients than the previous workshop form.

In 1960, Arthur Samuel (pioneer of artificial intelligence research) wrote the following, "Programming computers to play games is but one stage in the development of an understanding of the methods which must be employed for the machine simulation of intellectual behavior."[27] Among this early research was on Game Theory. Two areas of interest were:

- **Two-Player Perfect-Information Games**—"The most visible success for computer game-playing occurred in May 1997 when the chess machine DEEP BLUE defeated World Chess Champion Garry Kasparov in an exhibition match. This was a major success for artificial intelligence and a milestone in computing history."
- **Imperfect-Information and Stochastic Games**—"In some games, information is hidden (e.g., you cannot see the opponent's cards) and the program has to account for all possible scenarios, only one of which is actually valid." Backgammon and poker are examples of this game."[28]

There is a large body of work in this field. I have recommended to those looking at diversity that they review some of this as human are a very diverse and global species.

6.6.1 Non-Cooperative Games

Section 1.12.2.2 'Game Theory' of Chapter 1 is an overview of Game Theory and how it might play a role in AI-enabled training solutions. One game that is the predominate theory used by the Cross-Cultural Serious Game is the non-cooperative game briefly described herein.

> A non-cooperative game is one in which players are unable to make enforceable contracts outside of those specifically modeled in the game. Hence, it is not defined as games in which players do not cooperate, but as games in which any cooperation must be self-enforcing. Games in which players can enforce contracts through outside parties are termed cooperative games.[29]

6.6.1.1 The Cross-Cultural Game

Serious games are immersive (typically online) video/virtual reality games focused problem-solving rather than entertainment. This procedural simulation learning medium can help learners develop and hone complex competencies, i.e., aircraft pilot simulation, spacecraft training, among others.[30] One of the best features of a serious game is the ability to hit the reset button in the event of failure. No harm no foul training solution that we believe AI can dramatically enhance.

6.6.1.1.1 *Origin of the Game*

As the result of many years selling and licensing compute technology to global clients, it was necessary to develop a business model that enabled this sales representative and later sales executive to develop sustained business relationships with a large number of individuals from different countries and cultures. We developed this game from my 1996 doctoral dissertation as the result of a business need I had to better understand a large number of global cultures (the vast majority from Asia) quickly. At the time I was traveling globally, extensively with a major focus on Asia.

We have used this training tool since the mid-1990s with excellent results as its scenarios closely mirror real business issues. Key scenarios include Multi-Culture Team Negotiations, Industrial Safety Culture Interactions and Cross-Cultural Sales as well as custom scenarios.

The roots of the game are person to person or team to team using the Cloud to facilitate human interaction based on behavioral economics. With the maturity of AI, major new vistas are open for complex interaction training, including levels of difficulty step for increasingly sophisticated players.

6.6.1.1.2 *Rules of the Game*

The following is largely·taken from the Disparate Team Scenario game. These rules are basically the same for other scenarios, although some custom scenarios may vary.

Both teams should reach agreement in each of the three categories; however, each can have a different value. Moreover, the team is required reach agreement and offers all three metrics at the same time.

You can use any or all of the information in this scenario position paper during your interaction with the other player. Moreover, you can use your personal experience as well as other thoughts to craft your strategy.

Finally, you can exchange any information during this process, including 'miss-information' and even lie although you cannot reveal your sheet of performance metrics.

Don't forget to keep in mind the Organizational Traits and the Team Key Performance Indicators (KPIs) Your Team Selected in the Pre-Game Survey.

These three of the Performance Metrics is the same for both Players.

- **Performance**—The extent to which the team meets its success goals, measured by the KPIs each team has given
- **Cost**—This is the cost to the new (merged) company for completing the successful merger. It is measured not just in monetary costs but other metrics such as revenue grow/lost, turnover in the acquired firm, social impact, and other KPIs each team has been given
- **Trust**—The willingness of each organization to work together going forward as the new company and not the old one with a new division. It is measured by KPIs and other metrics in the game at the personal level

Each Team's Goal is to Maximize Its Set of Performance Metrics.

The term, "Maximize their Set or Portfolio of Performance Metrics" means to obtain the Best Overall. It does NOT mean that the Team is to necessarily attain the highest number in all categories but that the Sum of the three is the highest.

Economists call this the Efficiency Frontier (Section 6.6.2.3 'Pareto Optimality' of Chapter 6) because the set of optimal portfolios (three in this case) with the highest expected return for a defined level of risk. Other portfolios are sub-optimal where the risk-reward curve is higher. In other words, the risk is too high for the rate of return expected.

Note: this is generally not available until and number of teams have played this game, and their results reported.

As stated, the goal of each team is Maximize Its Set or Portfolio of Performance Metrics. Accomplishing this will lead to greater organizational performance and individual rewards such as bonuses and promotions, not to mention job satisfaction and overall employee morale.

6.6.1.1.3   Playing the Game

This is an online game. After each player logs in, they will be directed to complete a series of questions designed to identify demographics and various behaviors about you as an individual (or team) and their organization. Then each player will be provided with a scenario and set of rewards.

While the intent is to collaborate to solve a problem, the reward system builds in a level of tension. After the personal interaction, a post-game questionnaire is to be completed and finally, an analysis will be provided both sides.

6.6.1.1.3.1 Goals and Learning Experiences While we are using technology to facilitate this game, in reality this game is between people. Whether playing this game with another player or part of a team collaborating with the counterpart team, the goal is the same. Players can expect to gain:

- A greater understanding of cross-cultural interaction
- Empathy for counterpart(s)
- Specific expertise you can use immediately solving business problems

The game payoff matrix has three categories: Performance, Cost, and Trust. Each category has ten payoff values that are independently selected over a set of iterations until both parties agree to the deal.

By the Way, each player/team has a different list of payoff value unique to their role in the scenario. For example, levels of *Satisfaction* may be different.

6.6.1.1.4 Outcomes of the Game

The game generates a number of outcomes, many of which are difficult to accurately quantify. However, most will understand their meaning. Again, this is an area where AI can help assess these latent variables. Game Outputs include[31]:

Performance Criteria include:

- **Overall Assessment**—This assessment provides Players and management with a view of the overall outcome of the game and may identify areas that should be addressed. Specific recommendations to management are made including actionable items to increase the understanding between parties.
- **Payoff**—The goal of each player is Maximize their Set or Portfolio of Performance Metrics. Accomplishing this will lead to greater organizational performance and individual rewards such as bonuses and promotions.
- **Satisfaction**—This is the level of Satisfaction each Team has about the outcome of the game. This metric is a function the parties' Relationship.
- **Perceived Time**—Each Team is asked to provide the amount of Time the actual game required. This can be important in that negotiation time may appear to be short when 'things' are going well and drag when they are not.
- **Trust**—This is the level of Trust the Team has about the other Team.
- **Demographics**—Basic Demographics of each Player. Details are sanitized for privacy.
- **Post Game Team Attributes**—Each Player completes a Post Game Questionnaire. Effectively a set of questions about their perspective about their performance and the performance of the other Team. An average is taken for each Team.
- **Organizational Traits & High Reliability**—Each Organization has a set of Traits that will partially define its Culture. Each Player is asked for their 'perception' of these traits. This includes the aspects of the High Reliability mindset.

- **Culture of Safety Maturity**—Originally developed to meet the needs of heavy industry, Safety Culture Maturity models are extended to meet general workplace safety maturity.
- **KPIs**—According to Investopedia,

  > Key performance indicators (KPIs) refer to a set of quantifiable measurements used to gauge a company's overall long-term performance. KPIs specifically help determine a company's strategic, financial, and operational achievements, especially compared to those of other businesses within the same sector.[32]

- **Player Personality (Aggregate to Team)**—Often defined as behaviors, cognitions, and emotional patterns that evolve from biological and environmental factors.
  - **Agreeableness**—level of concern for social harmony
  - **Conscientiousness**—level of discipline v flexibility
  - **Extraversion**—level of breadth of activities not depth
  - **Neuroticism**—level of tendency to express negative emotions
  - **Openness**—level of willingness to try new
- **Player Temperament (Aggregate to Team)**—Often defined as an individual's 'human nature' from one perspective it consists of five elements:
  - **Choleric**—level of extroversion
  - **Melancholic**—level of analytical thinking
  - **Phlegmatic**—how laid back is the person
  - **Sanguine**—level of communicative behavior
  - **Supine**—level of need for inclusion
- **Relationships, Behaviors, Conditions (RBC)**
  - **Situational Awareness**—Generally defined as the subjective view of the setting one finds themselves in. It is generally seen as a critical element in the decision-making process.
  - The **Conditions** model includes the Circumstances, Culture, and Environment impacting on Behaviors and Relationship. Moreover, each Team's Capability is a function of Conditions as well. This can be construed as an understanding of one's situation aka Situational Awareness.
  - **Behaviors**—There are three elements to human interaction with this game; Organizational (see Org Traits), Team, and the Individual (Player).
  - **Relationships**—There are four levels: between the organizations as shown in Payoff, Satisfaction and Trust. Certainly, the last two are a function of Inter-organization, between Teams as well as Individuals in each Team and across Organizational boundaries.

There are a lot of variables that are difficult to measure and typically are only compared to the other party and in general as aggregated data. We believe this makes this game a very good candidate for AI.

Finally, one of the issue in international business, political or even social engagements is language. AI is enabling each player to read and respond in his or her native tongue.

6.6.1.1.5   The AI-Enabled New Version

We believe that 'Role-Based Prompting' (RBP) will keep the game scenario as real as it currently is in our human-only domain. This is a technique where AI is instructed to adopt a defined role, persons, or expertise. The value of RBP includes[33]:

- **Enhancing Relevance**—Responses tailored to the context of the role.
- **Improving Creativity**—Encourages AI to think outside of its default parameters.
- **Simply Complex Tasks**—"Assigning a role helps AI break down tasks into manageable steps."
- **Generate Diverse Outputs**—Different roles yield different perspectives and solutions.

Initially, we will develop a single AI version, meaning AI will only respond with a single response. Later if appropriate, the AI response will be more indicative of a team of different AI roles—just like human teams.

For example, the role and prompt might be:

- **Role**—Team Member
- **Prompt**—As a member of the team, respond to the latest proposal from the counterpart team.

Next version roles may include Team leader, Subject Matter Expert, Legal, etc. The intent is to populate the AI team with behaviors one might expect from human with those roles.

As noted in Section 2.3 'Relationships, Behaviors, and Conditions (RBC)' of Chapter 2, Relationships cannot be measure directly. We know the situation or at least something about it and we can see Behavior. We will treat Relationships as a set of latent variables in a similar manner as has been described herein.

As noted previously, the number of permutations that can be generated by each AI player/team is 10! (10 factorial). This slightly more that 3.6 million. Therefore, boundaries will need to be established as many will probably not make any sense (Figure 6.8).

Similar to playing chess, negotiation processes are iterative and can take some time to arrive at a deal. Moreover, people often change their minds (sometimes often), and a satisfactory conclusion is not guaranteed. This is the value of this training and comes very close to simulating real world environments.

6.6.1.1.6   AI as the Bully?

One possible scenario that humans may shy away from but is perhaps well suited for AI is the role of the bully. An unfeeling machine can play this part well. This type of scenario can be tailored for individuals of all ages, genders, and cultures and can simulate dealing with difficult people and challenging conflict Management.

We are going to look into this scenario using AI as part of the game's development process. Likely, this will not be a collaborative scenario.

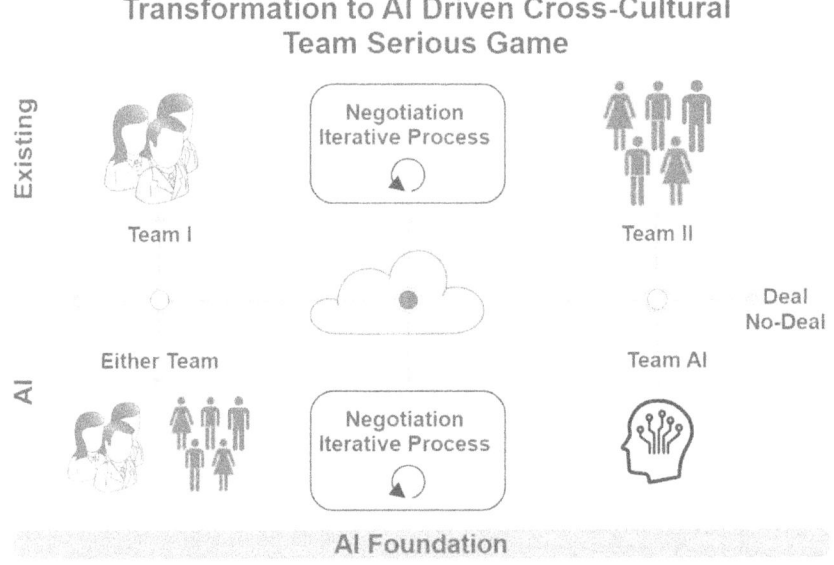

FIGURE 6.8   Transformation to AI-driven cross-cultural team serious game.

6.6.2   Case Six—Systemic Safety Index

Following the Deepwater Horizon catastrophe in April 2010, it was clear that significant changes in the Safety Culture of the upstream oil and gas industry had to be made. All US organizations are bound by OSHA regulations. However, Safety Culture is defined as:

> **"The core values and behaviors resulting from a collective commitment by leaders and individuals to emphasize safety, over competing goals, to ensure protection of people and the environment."**[34]

Moreover, organizations have allowed the appearance that safety a cost and hence they try to keep those expenditures as low as possible. What was needed is a definitive method that realistically correlates organizational performance with its safety record vis-à-vis its competitive peers as well as its position in its industry.

Since perception is reality, firms that desire to support a strong Safety Culture need metrics to convince a skeptical market and possibly management that their efforts are bearing fruit. What was needed is a tool that documents and demonstrates that high-performance organization truly have the best Safety Cultures.

Decades of data from Investor's Business Daily (IBD) have shown that firms with a Return on Equity (ROE) 17% or higher suggests that well-run companies outperform poorly managed ones. Major leaders often have ROE, or the annual net income divided by the average shareholder equity over the last two years from 25% to 50%.[35]

Over a decade ago, we made the case that catastrophic organizational failures, often resulting in loss of life, destroy shareholder value. Our 2014 book, Implementing a Culture of Safety: A Roadmap for Performance Based was the first usable set of implementation guidelines to develop and sustain a strong Safety Culture. Originally, written in response to the Deepwater Horizon disaster, it has found value in other critical sectors such as nuclear power generation. This roadmap also provides organizations with a self-administered Safety Cultural Maturity assessment program.[36]

As a follow-up on to this effort, we developed our Systemic Safet Index of Organizational Health. To our knowledge, it is the only model that addresses the complete range of variables that determine operational excellence as a function of systemic safety.

6.6.2.1 Background

This overview is adapted from our website and other internal company documents. We have developed a substantial body of knowledge on this topic that we continue to develop and feed into the model.

Typically, organizations and their industry measure against OSHA metrics and as well as Safety and Environmental Management Systems (SEMS) requirements where appropriate. These requirements continue to expand globally, albeit in somewhat different formats in some cases.

Other recent examples include Asset Integrity Management, i.e., ISO 55000. Moreover, High Reliability Management (HRM) and Human Factor Engineering (HFE), taken from other sectors are also finding favor.

However, these metrics are narrow and focus only on tasks performed by field personnel. What was needed is a more robust systemic model of organizational health and ability to implement and sustain high-level safety and environmental stewardship.

6.6.2.2 Organizational Health

Our organizational health matrix model specifically focuses on four components that directly affect an organization's ability to develop and sustain a robust Culture of Safety. This framework has two major axes, **Strategic** (vertical) or those foundational components that assure an organization is sustainable. For example, the financial strengths necessary as well as regulatory compliance that assures License to Operate.

From an **Operations** (horizonal) perspective, organizations require a high level of capability, including expertise on the part of their workforce as well as supply chain partners. Moreover, a systematic approach to risk management is probably the most challenging.

As shown in Figure 6.9, these four metrics form the basis of a comparative SSI Rating. Similar to financial metrics such as ROI, EPS, etc., this model enables those charged with Safety and Environmental management to assess the relative capability of an organization (such as its potential suppliers) to deliver sustainable performance. In some cases, over 100 variables are evaluated and a comparative

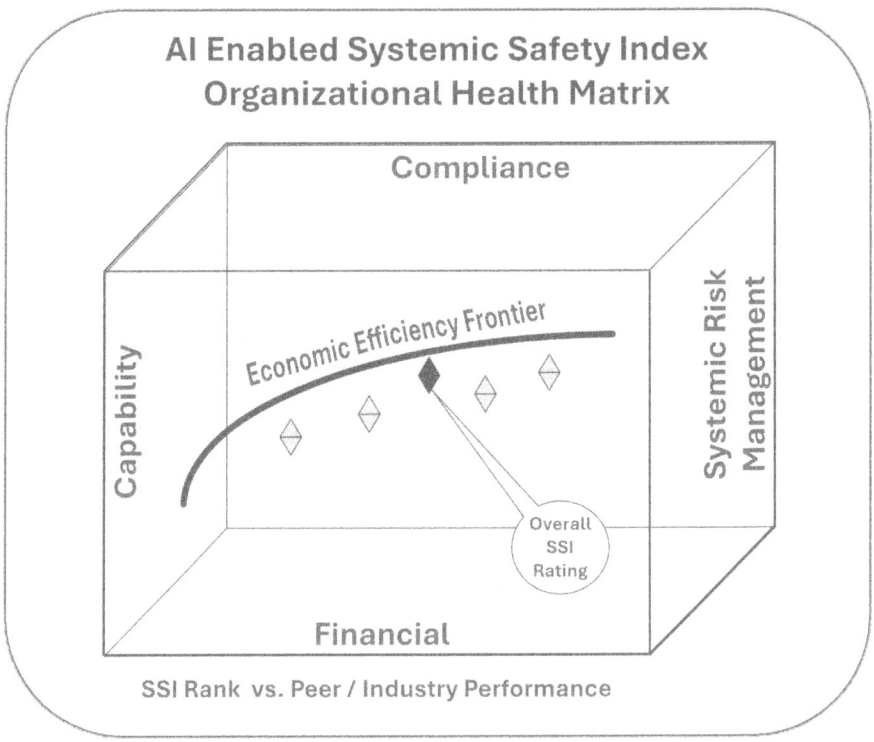

FIGURE 6.9 AI-enabled systemic safety index: Organizational health matrix.

weighted average ranking for each organization is provided from these four Health areas.

- **Financial**—All organizations seek to operate in a safe and environmentally responsible manner. However, some may not have the same level of capability to make the necessary investments in operations and training. Metrics such as Free Cash Flow, Depreciation and Working Capital provide insight into the financial ability to fund safety and environmental initiatives.
- **Compliance**—This is an assessment of the organization's level of compliance with relevant regulations and accepted industry standards. This metric and its moving average provide insight into the level of tolerance management has for accepting regulatory citations and fines.
- **Capability**—In addition to managerial expertise and turnover at the top levels, the KSA of engineering and field personnel is critical as well. Moreover, the quality of project management can be measured as a function of project delays and cost overruns that may sometimes lead to short cuts and other less than optimal performance.

- **Systemic Risk Management**—Robust Enterprise Risk Management (ERM) requires a holistic assessment of exposures to which the organization is exposed and mitigation processes put in place. One often-overlooked process is the quality of the OMS and its integration with the supply chain (Section 3.2.1 'Operations Management System' of Chapter 3).

Comparative Analysis

Each organization is measured against the overall industry practices. The matrix portfolio depicts where any given organization ranks against the industry's highest expected performance at levels of risk such as depicted in the following graphic.

The systemic safety index (SSI) assesses financial metrics, one index for operators and one for supply chain by region generic statistical study vs private report on any given company—peer group assessment for individual companies. A partial list of variable categories includes:

1. R&D
2. Return on Capital Employed (ROEC) and other relevant metrics
3. Depreciation
4. Free cash flow
5. Working capital
6. Quality of OMS
7. Regulatory violations
8. Fines
9. Project costs over run/time
10. Work force demographics
11. Confidential surveys
12. Human factor assessment
13. Systemic perspective
14. Typical workplace safety statistics
15. Complexity of corporate structure
    - Level of diversification
    - Definition of core business
    - Level of organizational chaos
    - Reorganization. M&A that focus by board and management
    - AEIG framework[37]
16. Culture—Rank order against cooperative vs command and control
17. Adherence to standards. For example, ISO standards
18. Reserve replacement rate
19. Strength of the Balance Sheet
20. Risk profile
21. Debt/equity ratio and other similar functions of liquidity
22. Data from states as well as federal/international sources
23. Restatements and one-time events
24. Use of tools such as Bowtie

As one might expect, this has been an arduous and time-consuming process. We expect the use of AI to dramatically impact on effectiveness and efficiency capturing and analyzing all these data.

6.6.2.3 Pareto Optimality

The goal is not to be the best in all categories. This would be an impossible task. The model is not an assessment of quality metrics such as the Malcolm Baldrige competition.[38]

Rather, we seek to better understand the organization in question's status without the preparation for an award. As part of a vendor selection or annual review process, all variables must meet *acceptable thresholds* relevant for said variable.

Only then do we drive toward Pareto optimality or the Economic Efficiency Frontier. As shown in the figure, we compare organizations against their peers and the industry in general.

Initially, these were custom assessments at the request of a client. They had a significant consulting component that was timely and expenses. Recently, we revamped that business model.

6.6.2.4 AI Enabled

One of the major challenges when assessing a large number of variables and comparing to other data is the sheer volume of effort required. This has limited this model. However, with the advent of useable AI, things have changed.

This type of analysis can extend to similar comparison such as equity performance and if coupled with the commodity future price influence model (Section 6.4) may significantly change the way financial and other analysts assess public company business performance and future expectations. We continue to develop and extend this model and will report further in future publications.

As with other behavioral models presented herein, we are using the same approach and technologies to help clarity latent variables.

NOTES

1. "5 Inspiring Walt Disney Quotes." Disney. https://news.disney.com/inspiring-walt-disney-quotes. Accessed October 9, 2025.
2. "Operations Management System: Smart OpEx." The Rapid Response institute. https://therrinstitute.com/operations-management-system/. Access July 31, 2025.
3. "AI Service Level Agreement (SLA) Generator." Sergei Tokmakov. https://terms.law/2024/10/25/ai-service-level-agreement-sla-generator/. Accessed August 1, 2025.
4. "Process Audit-Musk Style: A Systems Analysis Solution." The Rapid Response Institute. https://therrinstitute.com/process-audit-musk-style-systems-analysis-solution/. Accessed June 18, 2025.
5. "What Is Auditing?" ASQ. https://asq.org/quality-resources/auditing. Accessed June 18, 2025.
6. "DOGE Bypasses the Bureaucrats by Going Straight to the Payment Trails in the Government Computers." The Wentworth Report. https://wentworthreport.com/2025/02/09/doge-bypasses-the-bureaucrats-by-going-straight-to-the-payment-trails-in-the-government-computers/. Accessed September 7, 2025.

7. "NFL Mock Draft 2025: AI Predicts the Entire First Round." Honest AI. https://honestaiengine.com/nfl-mock-draft-2025-ai-predicts-the-entire-first-round?utm_source=ainews.honestaiengine.com&utm_medium=newsletter&utm_campaign=ai-draft-predictions-nfl-s-future-revealed&_bhlid=cd67a46d3579e869601df4c303363aa8b1897e54. Accessed March 30, 2025.

8. "NFL Draft Picks 2025: Live Results from Rounds 1-7." The Sporting News. https://www.sportingnews.com/us/nfl/news/nfl-draft-picks-2025-live-results/1df59786e708218d8793ee67. Accessed April 24, 2025.

9. "Shedeur Sanders Recent Slide Resurfaces His Damning Pre-Draft Process." Forbes. https://www.forbes.com/sites/kambuibomani/2025/04/26/shedeur-sanders-recent-slide-resurfaces-his-damning-pre-draft-process/. Accessed April 2025.

10. "AI Predicted the Next Pope. Did It Get It Right?" Science. https://www.science.org/content/article/ai-predicted-next-pope-did-it-get-it-right?utm_source=sfmc&utm_medium=email&utm_campaign=ScienceAdviser&utm_content=distillation&et_rid=1118318262&et_cid=5611078. Accessed May 9, 2025.

11. "Statistical Approaches for Small Numbers: Addressing Reliability and Disclosure Risk." NAHDO. https://www.nahdo.org/sites/default/files/publications/Data_Release_Guidelines.pdf. Accessed June 18, 2025.

12. "Rapid Response Management: Thriving in the New World Order." The Rapid Response Institute. https://therrinstitute.com/wp-content/uploads/2017/10/rapid_response_management_-_thriving_in_the_new_world_order_2.0_-_january_2009.pdf. Accessed August 1, 2025.

13. "Oil (and/or Gas) Reserve Definitions." BP. https://www.bp.com/content/dam/bp/business-sites/en/global/corporate/pdfs/energy-economics/statistical-review/bp-stats-review-2021-oil-reserve-definitions.pdf. Accessed March 11, 2025.

14. "Top Oil Executives Reckon with Downturn Even as Trump Cheers Them On." Reuters. https://finance.yahoo.com/news/top-oil-executives-reckon-downturn-005734370.html. Accessed March 10, 2025.

15. "Energy & Financial Markets: What Drives Crude Oil Prices?" EIA. https://www.eia.gov/finance/markets/crudeoil/. Accessed August 1, 2025.

16. "function." Britannica. https://www.britannica.com/science/function-mathematics. Accessed July 29, 2025.

17. "Latin Small Letter F." https://wumbo.net/symbols/f/. Accessed October 3, 2025.

18. "The Synergy of Hard and Soft Metrics In Decision-Making." Forbes. https://www.forbes.com/councils/forbesbusinesscouncil/2023/10/06/the-synergy-of-hard-and-soft-metrics-in-decision-making/#:~:text=Hard%20metrics%20are%20the%20data%20and%20charts%20you,social%20media%20say%20positive%20things%20about%20your%20company. Accessed March 10, 2025.

19. "Y as a Function of X Examples – Understanding Mathematical Relationships." Story of Mathematics. https://www.storyofmathematics.com/y-as-a-function-of-x-examples/. Accessed March 12, 2025.

20. Kuiper, Marcus A. and Shemwell, Scott M. (2013, February). "Mitigating Operational Risk Using the Power of Social Media." Petroleum Africa Magazine. pp. 28–31.

21. "AI Techniques Excel at Solving Complex Equations in Physics, Especially Inverse Problems." PHYS.ORG. https://phys.org/news/2025-10-ai-techniques-excel-complex-equations.html. Accessed October 3, 2025.

22. Shemwell, Scott M. (2025). Navigating the Data Minefields: Management's Guide to Better Decision-Making. CRC Press/Taylor & Francis. pp. 12, 13.

23. "What Is Agentic AI, and How Will It Change Work?" Harvard Business Review. https://hbr.org/2024/12/what-is-agentic-ai-and-how-will-it-change-work. Accesses March 31, 2025.

24. "Welcome to Our Cross-Cultural Serious Game Portal." The Rapid Response Institute. https://therrinstitute.com/lets-play-a-game/. Accessed March 31, 2025.

25. "Is Getting an AI Agent for Your Machine Tools a Good Idea?" IndustryWeek. https://www.industryweek.com/technology-and-iiot/article/55277876/is-getting-an-ai-agent-for-your-machine-tools-a-good-idea?o_eid=7584G5939223H0L&oly_enc_id=7584G5939223H0L&rdx.ident[pull]=omedal7584G5939223H0L&utm_campaign=CPS250326240&utm_medium=email&utm_source=IY+IW+Daily+Headlines+-+Morning. Accessed March 31, 2025.
26. "Using AI in Predictive Maintenance." Oracle. https://www.oracle.com/scm/ai-predictive-maintenance/. Accessed October 3, 2025.
27. A. Samuel, Programming Computers to Play Games, in: F. Alt (Ed.), Advances in Computers, Vol. 1, Academic Press, New York, 1960, pp. 165–192.
28. "Games, Computers, and Artificial Intelligence. Elsevier. PII: S0004-3702(01)00165-5. Accessed July 1, 2025.
29. "Non-Cooperative Game." GameTheory.net. https://www.gametheory.net/dictionary/Non-CooperativeGame.html. Accessed July 3, 2025.
30. "Serious Games." ScienceDirect. https://www.sciencedirect.com/topics/engineering/serious-games. Accessed July 1, 2025.
31. "Demonstration Game Results." The Rapid Response Institute. https://therrinstitute.com/diverse-teams-game-analysis/. Accessed October 3, 2025.
32. "KPIs: What Are Key Performance Indicators? Types and Examples." Investopedia. https://www.investopedia.com/terms/k/kpi.asp. Accessed April 5, 2025.
33. "Mastering Role-Based Prompting: The Secret to Smarter AI Interaction." SP Cloud Academy. https://spca.education/mastering-role-based-prompting-the-secret-to-smarter-ai-interactions/. Accessed October 3, 2025.
34. Holland, Winford "Dutch" E. and Shemwell, Scott M. (2014). Implementing a Culture of Safety: A Roadmap to Performance-Based Compliance. Xlibris, New York.
35. "Well-Managed Companies Boast Stellar Return on Equity." Investor's Business Daily. https://www.investors.com/how-to-invest/investors-corner/high-return-on-equity-sign-of-well-run-company/. Accessed June 10, 2025.
36. Holland, Winford "Dutch" E. and Shemwell, Scott M. (2014). Implementing a Culture of Safety: A Roadmap to Performance-Based Compliance. Xlibris, New York.
37. "Asset/Equipment Integrity Governance: Operations-Enterprise Alignment. The Rapid Response Institute. https://therrinstitute.com/wp-content/uploads/2017/10/asset_integrity_governance_-ver_1.1.pdf. Accessed October 3, 2025.
38. "What Is the Malcolm Baldrige National Quality Award (MBNQA)?" ASQ. https://asq.org/quality-resources/malcolm-baldrige-national-quality-award. Accessed June 10, 2025.

7 Conclusion and Way Forward

It is gratifying for this physicist whose degree was once looked down upon by hiring managers when I was a new entrant to the workforce as not being engineering to see AI enthusiastically embrace mathematics and construct of physics. Given the nature of the technologies and the tasks they will solve, it is appropriate that the "study of the underlying laws and mechanisms explaining how the universe works," take such a leadership and fundamental role in Artificial Intelligence.[1]

7.1 POINT OF NO RETURN

In physics, an Event Horizon (EH) is the boundary between a black hole and the rest of the universe. This is an important construct in that once something passes the EH, it cannot return to our 'normal' reality. Within the EH is the black hole's Singularity, a point where the laws of physics break down and the fabric of space and time curve an infinite degree.[2] This is definition of the 'point of no return.'

In June 2025, Sam Altman, the CEO of OpenAI proclaimed that humanity had already crosses an Event Horizon or Critical Inflection Point; the era of digital Superintelligence had begun. He defined what he called a 'gentle singularity' or a manageable transition to the unknowns of the new superintelligence reality. He went on to indicate even small misalignments of technology delivering benefits to humanity could have disastrous consequences. He suggested, as discussed in this work an approach that included:

- Focus on the long-term direction and goals of humanity, not just the short-term quick returns on cool technologies.
- Avoid the concentration of AI power in one individual, entity and even country or region of the world.
- The ethics, values, and limitations need to be established as the guidelines going forward.[3]

There have been a lot of monumental changes over the last 100 years. The closest parallel to superintelligence may be the dawn of *dual-use* technology (weaponized/useful) nuclear power in 1945. Most readers are well aware of the checkered past of nuclear as well as current concerns regarding proliferation and climate/radiation exposures.

Initially a weapon of war, nuclear power has faced societal and ethical issues throughout its existence. In its current incarnation this source of almost endless electrical generation may be the technological savior allowing the next-generation AI to exceed the wildest dreams of AI postulates. To date, nuclear technologies have been managed and secured reasonably well in that the last bomb was dropped in 1945 and the damage to facilities is largely minimal at the global, macro-level.

Benefits have exceeded the negative only because ethical models have been followed and many bad global actors have been prevented from attaining 'the bomb.' It is the closest good/actual practice society has used for existential technologies. This is a good start for the management of AI.

Others may disagree and that is a good thing. There are no simple issues with superintelligence, and a broad serious ongoing discussion is necessary.

7.2   AI TOYS

Most parents agree that play is an important part of children's development. Most would also agree that the desire to play continues throughout the human life cycle. The same is probably true for domestic and wild animals as well, i.e., dogs, elephants, dolphins, and others.

We could expect that AI will play a role in this process. We have already seen how AI drives games, both serious and entertainment. In June 2025, OpenAI and the toy maker Mattel announced an agreement to jointly develop an age-appropriate 'play experience' with an emphasis on innovation, privacy, and most of all safety. Not only enriching the play experience, AI-powered toys can provide valuable experience using this technology that can transcend into adulthood.[4] Perhaps ultimately the way basics such as reading, writing, and arithmetic prepare kids for social engagement.

We can expect AI to play an increasing role in education at all levels in the years to come. Several years ago, I met a father who explained that he had been very concerned that his then teenage boy was excessively into video game. He wondered if that time of his youth was just wasted, that it until his adult child landed a job 'driving' remote underwater vehicles. His ability to live in the virtual 3D space enabled him to be successful in a career path he loved. Some drones required the same skill set. Perhaps this is the future for many AI kids.

7.3   SOCIETAL GOOD, BAD, AND UGLY

Human behavior covers the gamut. We are capable of enormous positive value such as medical discoveries as well-unfettered evil and use all the tools at our disposal to accomplish both.

7.3.1   INDIVIDUAL SAFETY

We are making great strides in the use of AI to detect threats and keep us safe. TSA has announced that they will use facial recognition. The following two deployed projects are indicative of how AI is being used.[5] With over ten new AI-enabled safety project underway, one can expect the use of AI for travel safety to increase dramatically.

- **Low Probability of False Alarm (Low-PFA) Algorithm for On-Person Screening**—"TSA uses Low Probability of False Alarm (Low-PFA) algorithms to train AI systems deployed for on-person screening. It utilizes ML to improve detection performance while decreasing alarm rates and passengers touch rates."

- **Airport Throughput Predictive Model**—"This use case is a predictive model for passenger volume to help with airport staffing. The Airport Throughput Predictive Model is an AI model which ingests checkpoint screening throughput data to train the predictive model, to provide projections for future-date throughput."

One of the major current political issues is immigration or migration. Some think that unlimited access is the right of every individual while others are appalled by the levels of the immediate past. US elected individuals and it seems other country political appear to run from the issue. AI technology plays a role in this process but only raises a specific individual issue. Does an identified person need to be deported? This is a legal and social issue, and AI plays a major role in the initial identification process.[6]

This is but one example of a controversial use of AI. This is likely an ongoing issue with legs similar to concerns over social media.

7.4   SPACE IS THE LIMIT?

Physicists know that *space* can be viewed as infinitely large and infinitely small. In this book we sought to help non-AI individuals develop an understanding of the technology, where, when, and how to use it as well as its risks and limitations.

NASA is using the concept of Dynamic Targeting to make satellites 'think' more like humans reasoning that "enable spacecraft to decide, autonomously and within seconds, where to best make science observations from orbit."

> The technology enabled an Earth-observing satellite for the first time to look ahead along its orbital path, rapidly process, and analyze imagery with onboard AI, and determine where to point an instrument. The whole process took less than 90 seconds, without any human involvement.[7]

This AI solution can have broad use across all sensor processes and is a potential game-changer by itself.

The AI era has begun and will accelerate. Vested interests of all natures have a so-called AI dog in the fight and massive amounts of capital are already committed. Likened to the next Industrial Revolution, it will soon be as ubiquitous as the automobile.

7.5   FINAL THOUGHTS

One of the major challenges of a book of this nature is the fast-moving nature of the subject. We attempted to provide thoughts that are more timeless than the rollout of the next AI version.

Our work-in-progress case studies describe processes that are typically not featured in colloquial media and our discussion regarding intelligence, prompting and structural equations cast light on those projects that are very complicated. The true value from AI is not in chatbots or simply doing one's homework, rather it will be found solving the world's toughest problems.

The challenge then will be to make sure that AI accurately solved the problem as bias free as possible. I am concerned that this process still has many gaps. Moreover, like all technologies, how does society protect its children from those who would use the technology for nefarious purposes or biases that support destructive behaviors?

In this book we referenced materials using The Chicago Manual of Style and other tools to assure a high quality of the work. Perhaps, AI needs a style guide that supports the research by citing its origin (and a date on the source would be most helpful).

To some extent, a few AI inference engines provide these linkages. This is a step-up from the wild west days of the Internet where copyright were routinely overlooked or violated.

Much of this has yet to be developed. The greater challenge will not be technology but how to capitalize on for the betterment of humankind or not.

I want to thank you for reading this book and wish you success on your Artificial Intelligence Journey. It should be quite a ride!

NOTES

1. "What Is Physics?" Michigan Tech. https://www.mtu.edu/physics/what/. Accessed June 12, 2025.
2. "What Is a Black Hole Event Horizon (and What Happens There)?" Space.com. https://www.space.com/black-holes-event-horizon-explained.html. Accessed June 14, 2025.
3. "OpenAI CEO Says We've Already Passed the 'Superintelligence Event Horizon'." Emerge. https://decrypt.co/324532/openai-ceo-says-weve-already-passed-the-superintelligence-event-horizon. Accessed June 14, 2025.
4. "Mattel and OpenAI Announce Strategic Collaboration." Mattel. https://corporate.mattel.com/news/mattel-and-openai-announce-strategic-collaboration?utm_source=newsletter.genai.works&utm_medium=newsletter&utm_campaign=openai-mattel-the-era-of-ai-toys-begins&_bhlid=71b77f03e68e79d676e8dacbfc15104dab878193. Accessed July 27, 2025.
5. "Transportation Security Administration – AI Use Cases." Homeland Security. https://www.dhs.gov/ai/use-case-inventory/tsa. Accessed October 3, 2025.
6. "How Immigrants and Protesters Are Being Caught in ICE's AI Dragnet." Emerge. https://decrypt.co/325003/how-immigrants-and-protesters-are-being-caught-in-ices-ai-dragnet. Accessed October 3, 2025.
7. "How NASA Is Testing AI to Make Earth-Observing Satellites Smarter." NASA. https://www.nasa.gov/science-research/earth-science/how-nasa-is-testing-ai-to-make-earth-observing-satellites-smarter/?utm_source=Generative_AI&utm_medium=Newsletter&utm_campaign=meta-names-ex-openai-scientist-to-head-superintelligence-labs&_bhlid=4173a4c020e532434df54d1d9ae3d301fe8888ff. Accessed July 28, 2025.

Appendix I
Glossary of Terms

The following terms are used in this report and by practitioners. These definitions are provided for the reader's convenience. They can also be used in Key Word searches. As a general rule, the following definitions are taken from the relevant web site and are references as appropriate. As used throughout this book, direct quoted text is in *italics*, per editorial convention.

TABLE A.1
Glossary of Terms

Term	Definition
Adoption Readiness Level (ARL) Framework	*A tool to assess the commercialization risks facing a technology as it crosses the Research, Development, Demonstration, and Deployment (RDD&D) continuum to reach successful commercialization. The ARL framework complements the widely adopted Technology Readiness Levels (TRL) framework by extending beyond the technical risks that technologies face on their commercialization journeys to capture other commercialization risks. It consists of 17 dimensions that fall into four risk buckets.*[1]
Advanced Persistent Threat (APT)	*A sophisticated, sustained cyberattack in which an intruder establishes an undetected presence in a network in order to steal sensitive data over a prolonged period of time. An APT attack is carefully planned and designed to infiltrate a specific organization, evade existing security measures and fly under the radar.*[2]
Agentic AI	*It refers to AI systems and models that can act autonomously to achieve goals without the need for constant human guidance. The agentic AI system understands what the goal or vision of the user is and the context to the problem they are trying to solve.*[3]
AI Detection	*Grounded in the principles of natural language processing (NLP) and machine learning, AI detection offers a robust framework for discerning AI-generated text from human-written works.*[4]
AI Ethics	*Ethics is a set of moral principles which help us discern between right and wrong. AI ethics is a multidisciplinary field that studies how to optimize the beneficial impact of artificial intelligence (AI) while reducing risks and adverse outcomes.*[5]
AI Hallucination	*A phenomenon wherein a large language model (LLM)—often a generative AI chatbot or computer vision tool—perceives patterns or objects that are nonexistent or imperceptible to human observers, creating outputs that are nonsensical or altogether inaccurate.*[6]

(Continued)

193

TABLE A.1 (*Continued*)
Glossary of Terms

Term	Definition
Algorithm	*In the context of computer science, an algorithm is a mathematical process for solving a problem using a finite number of steps. Algorithms are a key component of any computer program and are the driving force behind various systems and applications, such as navigation systems, search engines, and music streaming services.*[7]
Application Programming Interface (API)	*A set of commands, functions, protocols, and objects that programmers can use to create software or interact with an external system. It provides developers with standard commands for performing common operations, so they do not have to write the code from scratch.*[8]
Bayesian Optimization	*In Machine Learning, (it) is an optimization method that uses probabilistic models to efficiently find a model's hyperparameters. In other words, it's a mathematical technique whose objective is to find the optimal combination of hyperparameters.*[9]
Bias	*A bias is a tendency, inclination, or prejudice toward or against something or someone. Some biases are positive and helpful—like choosing to only eat foods that are considered healthy or staying away from someone who has knowingly caused harm. But biases are often based on stereotypes, rather than actual knowledge of an individual or circumstance. Whether positive or negative, such cognitive shortcuts can result in prejudgments that lead to rash decisions or discriminatory practices.*[10]
Bionics	*Science of constructing artificial systems that have some of the characteristics of living systems. Bionics is not a specialized science but an interscience discipline; it may be compared with cybernetics. Bionics and cybernetics have been called the two sides of the same coin. Both use models of living systems, bionics in order to find new ideas for useful artificial machines and systems, cybernetics to seek the explanation of living beings' behaviour.*[11]
Black Swan Event	*An extremely negative event or occurrence that is impossibly difficult to predict. In other words, black swan events are events that are unexpected and unknowable.*[12]
Blockchain Credentialing	*Blockchain credentialing is the ultimate innovation in education, delivering rock-solid security and transparency. It transforms credentials into unbreakable, easily accessible assets, fostering inclusivity and equal opportunities. Globally, blockchain-based credentials are the new power move, empowering individuals to effortlessly manage and showcase their verified achievements.*[13]
Bloom's Taxonomy	*A set of three hierarchical models used to classify educational learning objectives into levels of complexity and specificity. The three lists cover the learning objectives in cognitive, affective, and sensory domains, namely: thinking skills, emotional responses, and physical skills.*[14]
Brain-Computer Interface (BCI)	*A system comprising a direct communication pathway between the brain and an external device.*[15]
Chain of Thought	*Mirrors human reasoning, facilitating systematic problem-solving through a coherent series of logical deductions.*[16]

(*Continued*)

TABLE A.1 (*Continued*)
Glossary of Terms

Term	Definition
Checksum	*A value used to verify the integrity of a file or a data transfer. In other words, it is a sum that checks the validity of data. Checksums are typically used to compare two sets of data to make sure they are the same.* [17]
Context Window	*The context window (or 'context length') of a large language model (LLM) is the amount of text, in tokens, that the model can consider or 'remember' at any one time. A larger context window enables an AI model to process longer inputs and incorporate a greater amount of information into each output.* [18]
Contrastive Learning	*A type of unsupervised learning technique that aims to learn representations by contrasting positive pairs (i.e., similar samples) with negative pairs (i.e., dissimilar samples). This approach is based on the idea that by learning to distinguish between similar and dissimilar samples, a model can develop a robust understanding of the underlying structure of the data.* [19]
Critical Thinking	*The act or practice of thinking critically (as by applying reason and questioning assumptions) in order to solve problems, evaluate information, discern biases, etc. Today, what we call the **Socratic** method is a way of teaching that fosters critical thinking, in part by encouraging students to question their own unexamined beliefs, as well as the received wisdom of those around them.* [20]
Cross-Selling	*A sales strategy where a seller offers complementary products or services to existing customers, delivering more value while increasing revenue. It requires identifying additional customer needs and proposing offerings that meet those needs, increasing satisfaction and boosting sales.* [21]
Cultural Cognition	*The tendency of people to fit their perceptions of risk and related facts to their group commitments.* [22]
Culture	*The set of shared attitudes, values, goals, and practices that characterizes an institution or organization.* [23]
Curse of Dimensionality	*Coined by mathematician Richard E. Bellman, the curse of dimensionality references increasing data dimensions and its explosive tendencies. This phenomenon typically results in an increase in computational efforts required for its processing and analysis.* [24]
Dark Energy (in Physics)	*Repulsive force that is the dominant component (69.4 percent) of the universe. The remaining portion of the universe consists of ordinary matter and dark matter.* [25]
Dark Matter (in Physics)	*Dark matter constitutes over 80% of all matter in the universe, yet it remains unseen by scientists. Its existence is inferred because, without it, the behavior of stars, planets, and galaxies would be inexplicable.* [26]
Data, Multimodal	*Type of data that is collected from multiple sources, formats, and modalities.* [27]
Data Validation	*The practice of checking the integrity, accuracy and structure of data before it is used for or by one or more business operations. The results of a data validation operation can provide useful, actionable data that can then be used for data analytics or business intelligence applications, or for training machine learning models.* [28]

(*Continued*)

TABLE A.1 (*Continued*)
Glossary of Terms

Term	Definition
Deep Architectures	*Connectionist architectures have existed for more than 70 years, but new architectures and graphical processing units (GPUs) brought them to the forefront of artificial intelligence. Deep learning isn't a single approach but rather a class of algorithms and topologies that you can apply to a broad spectrum of problems.*[29]
Deep Neural Network	*It imitates the functioning of the human brain. Find out everything you need to know about it: definition, functioning, use cases, training. A neural network is a set of algorithms inspired by the human brain. The goal of this technology is to simulate the activity of the human brain, and more specifically the recognition of patterns and the transmission of information between the different layers of neural connections.*[30]
Detector Zoo	*The release of 50 top-performing designs in a public repository, dubbed the 'Detector Zoo,' exemplifies a movement towards open data initiatives.*[31]
Difference Equations	*Differential equation are great for modeling situations where there is a continually changing population or value. If the change happens incrementally rather than continuously then differential equations have their shortcomings. Instead, we will use difference equations which are recursively defined sequences. Examples of incremental changes include salmon population where the salmon spawn once a year, interest that is compound monthly, and seasonal businesses such as ski resorts.*[32]
Diffusion Model	*Generative models used primarily for image generation and other computer vision tasks. Diffusion-based neural networks are trained through deep learning to progressively 'diffuse' samples with random noise, then reverse that diffusion process to generate high-quality images.*[33]
Dormitive Principle	*A type of tautology in which an item is being explained in terms of the item itself, only put in different (usually more abstract) words.*[34]
Dunning–Kruger Effect	*A cognitive bias in which people with limited competence in a particular domain overestimate their abilities. It was first described by Justin Kruger and David Dunning in 1999. Some researchers also include the opposite effect for high performers: their tendency to underestimate their skills. In popular culture, the Dunning–Kruger effect is often misunderstood as a claim about general overconfidence of people with low intelligence instead of specific overconfidence of people unskilled at a particular task.*[35]
Dynamic Targeting	*The idea is to make the spacecraft act more like a human: Instead of just seeing data, it's thinking about what the data shows and how to respond. When a human sees a picture of trees burning, they understand it may indicate a forest fire, not just a collection of red and orange pixels. We're trying to make the spacecraft have the ability to say, 'That's a fire,' and then focus its sensors on the fire.*[36]
Eisenhower Matrix	*Also known as the Eisenhower Box or Urgent-Important Matrix, it is defined as a pivotal time management and productivity tool, designed to empower individuals in efficiently prioritizing tasks. This strategic framework categorizes tasks into four distinct quadrants—Do First, Schedule, Delegate, and Eliminate/Do Last—based on their urgency and importance.*[37]

(*Continued*)

TABLE A.1 (*Continued*)
Glossary of Terms

Term	Definition
Embeddings	*Continuous vector representations of discrete data. They serve as a bridge between the raw data and the machine learning models by converting categorical or text data into numerical form that models can process efficiently. The goal of embeddings is to capture the semantic meaning and relationships within the data in a way that similar items are closer together in the embedding space.*[38]
Expert System	*A computer program that uses artificial intelligence (AI) technologies to simulate the judgment and behavior of a human or an organization that has expertise and experience in a particular field.*
	Expert systems are usually intended to complement, not replace, human experts.[39]
	The concept of expert systems was developed in the 1970s by computer scientist Edward Feigenbaum, a computer science professor at Stanford University and founder of Stanford's Knowledge Systems Laboratory.
Fine-Tuning	*In machine learning, it is the process of adapting a pre-trained model for specific tasks or use cases. It has become a fundamental deep learning technique, particularly in the training process of foundation models used for generative AI. Fine-tuning could be considered a subset of the broader technique of transfer learning: the practice of leveraging knowledge an existing model has already learned as the starting point for learning new tasks.*[40]
Fear of Becoming Obsolete (FOBO)	*The workplace anxiety gripping employees who worry that AI, automation, and shifting job expectations are outpacing their skills.*[41]
Future Proofing	*Design or change it so that it will continue to be useful or successful in the future if the situation changes.*[42]
Generative Engine Optimization	*It is the process of ensuring your digital content maximizes its reach and visibility inside of Generative AI Engines like ChatGPT, Claude, SGE, Gemini, Perplexity, and more when people inquire about: Solutions and products you sell, Stories that you have been in, Services that you offer, Ideas you have shared, and Information in which you have deep expertise and experience.*[43]
Graphics Processing Unit (GPU)	*A specialized electronic circuit originally designed to speed up the creation of images and videos. However, its remarkable ability to perform vast numbers of calculations rapidly has led to its adoption in diverse fields, including artificial intelligence and scientific computing, where it excels at handling data-intensive and computationally demanding tasks.*[44]
Heuristics	*Involving or serving as an aid to learning, discovery, or problem-solving by experimental and especially trial-and-error methods.*[45]
Independent Verification & Validation (IV&V)	*A comprehensive review, analysis, and testing (software and/or hardware) performed by an objective third party to confirm (i.e., verify) that the requirements are correctly defined, and to confirm (i.e., validate) that the system correctly implements the required functionality and security requirements.*[46]

(*Continued*)

TABLE A.1 (*Continued*)
Glossary of Terms

Term	Definition
Inference-as-a-Service	*An enterprise application* (that) *interfaces with the machine learning model (in this case, the LLM), with low operational overhead. This means you can run your code to interface with the LLM without focusing on infrastructure.*[47]
Inference Engine	*The brain of the ES* (Expert System). *It is also known as the control structure or rule interpreter (in rule-based ES). This component is essentially a program that provides a methodology for reasoning about information in the knowledge base and on the blackboard, and for formulating conclusions. This component provides directions about how to use the system's knowledge by developing the agenda that organizes and controls the steps taken to solve problems whenever consultation is performed.*[48]
Inference-Time Scaling	*Also known as test-time scaling or AI reasoning. It's a technique that allows AI models to take their time to 'think' instead of rushing towards an answer. It lets the models evaluate multiple solutions before picking the best one, much like how humans approach tough problems.*[49]
Instant or Immediate Gratification	*Refers to the temptation, and resulting tendency, to forego a future benefit in order to obtain a less rewarding but more immediate benefit.*[50]
Joint Embedding Predictive Architecture (JEPA)	*It is a self-supervised learning framework designed to learn useful representations from data without relying on labeled examples. JEPA is particularly notable for its focus on learning abstract, high level representations that are robust and generalize well across tasks.*[51]
Key Performance Indicators (KPIs)	*Measure a company's success versus a set of targets, objectives, or industry peers.*[52]
Knowledge Graph	*A structured, graph-based representation of entities and the relationships between them. In essence, a Knowledge Graph transforms disconnected data into actionable knowledge, enabling computers to 'think' and respond more intelligently, mirroring how our own brains connect ideas to comprehend the world around us. This approach enables both humans and machines to understand, reason about, and extract insights from complex data.*[53]
Knowledgeable Buyer	*Parties define knowledge so that the rules of the game are clear. Fundamentally, knowledge definitions seek to delineate whose knowledge matters for the purposes of determining whether a knowledge-qualified representation has been breached. The reason this is important is because, absent a contractual limitation, courts may be willing to impute knowledge to a pool of people that is larger than intended.*[54]
Large Language Model (LLM)	*A category of foundation models trained on immense amounts of data making them capable of understanding and generating natural language and other types of content to perform a wide range of tasks.*[55]
Large Reasoning Model (LRM)	*A new category of specialized language models. They are designed to break down complex problems into smaller, manageable steps and solve them through explicit logical reasoning (This step is also called 'thinking'). Unlike general-purpose LLMs which might generate direct answers, reasoning models are specifically trained to show their work and follow a more structured thought process.*[56]

(*Continued*)

TABLE A.1 (*Continued*)
Glossary of Terms

Term	Definition
Latent Variables	*Latent variables represent underlying factors or constructs that are not directly observable but are inferred from other measurable variables.*[57]
Law of Unintended Consequences	*Actions of people—and especially of government—always have effects that are unanticipated or unintended. Economists and other social scientists have heeded its power for centuries; for just as long, politicians and popular opinion have largely ignored it.* *The concept of unintended consequences is one of the building blocks of economics. Adam Smith's 'invisible hand,' the most famous metaphor in social science, is an example of a positive unintended consequence.*[58]
LISREL Model	*Causal modeling as defined and developed by Joreskog (1973, 1977, 1978) is known as the Linear Structural Relationship Model, or LISREL. LISREL consists of the structural equation model and the measurement model. The structural equation model describes the theoretical causal relationships among the latent variables via a set of general linear equations. The measurement model describes the measurement of the latent variables by the observable indicator variables and allows evaluation of the measurement properties of such measures.*[59]
Marginal Cost	*The additional cost incurred in the production of one more unit of a good or service. It is derived from the variable cost of production, given that fixed costs do not change as output changes, hence no additional fixed cost is incurred in producing another unit of a good or service once production has already started.*[60]
Meta Prompting	*An advanced prompt engineering technique where prompts are used to generate, refine, or analyze other prompts, rather than directly answering a user's question. This higher-level approach helps guide large language models (LLMs) to create, improve or interpret prompts for specific tasks, making AI interactions more dynamic, flexible and effective.*[61]
Model Collapse	*The declining performance of generative AI models that are trained on AI-generated content.*[62]
Model Context Protocol (MCP)	*Serves as a standardization layer for AI applications to communicate effectively with external services such as tools, databases and predefined templates.*[63]
Moral Foundations Theory (MFT)	*In essence, MFT suggests that there are several innate psychological systems at the core of our 'intuitive ethics.' Cultures then build virtues, narratives, and institutions upon these foundational systems, resulting in the diverse moral beliefs we observe globally and even conflicts within nations.*[64]
Morphogenesis	*The shaping of an organism by embryological processes of differentiation of cells, tissues, and organs and the development of organ systems according to the genetic 'blueprint' of the potential organism and environmental conditions.*[65]
Natural Language Processing (NLP)	*NLP enables computers and digital devices to recognize, understand and generate text and speech by combining computational linguistics, the rule-based modeling of human language together with statistical modeling, machine learning and deep learning.*[66]

(*Continued*)

TABLE A.1 *(Continued)*
Glossary of Terms

Term	Definition
North Star Metric	*Should reflect the core value you deliver to customers. It needs to be measurable, actionable, and a leading indicator of future success. By consistently optimizing for this metric, you'll drive meaningful growth and keep your customers happy.*[67]
Numerical Relativity	*A multidisciplinary field including relativity, magneto-hydrodynamics, astrophysics and computational methods, among others, with the aim of solving numerically highly-dynamical, strong-gravity scenarios where no other approximations are available.*[68]
Open Source	*Software source code that is made freely available for possible modification and redistribution. Products include permission to use the source code, design documents, or content of the product. The open-source model is a decentralized software development model that encourages open collaboration.*[69]
Pareto Efficiency or Pareto Optimality	*An economic state where resources cannot be reallocated to make one individual better off without making at least one individual worse off. Pareto efficiency implies that resources are allocated in the most economically efficient manner but does not imply equality or fairness.*[70]
Physical Symbol System	*Consists of a set of entities, called symbols, which are physical patterns that can occur as components of another type of entity called an expression or symbol structure.*[71]
Predictive AI	*Predictive artificial intelligence (AI) involves using statistical analysis and machine learning (ML) to identify patterns, anticipate behaviors and forecast upcoming events. Organizations use predictive AI to predict potential future outcomes, causation, risk exposure and more.*[72]
Prompt Chaining	*A natural language processing (NLP) technique, which leverages large language models (LLMs) that involves generating the desired output by following a series of prompts. In this process, a sequence of prompts is provided to an NLP model, guiding it to produce the desired response. The model learns to understand the context and relationships between the prompts, enabling it to generate coherent, consistent, and contextually rich text.*[73]
Prompt Engineering	*The practice of designing inputs for AI tools that will produce optimal outputs.*[74]
Prompt Library	*A collection of predesigned prompts, serving as templates to expedite the creation of AI prompts. This resource accelerates development, and ensures best practices are followed in prompt engineering.*[75]
Quantum Computing	*An emergent field of cutting-edge computer science harnessing the unique qualities of quantum mechanics to solve problems beyond the ability of even the most powerful classical computers.*[76]
Red Team	*A group of people authorized and organized to emulate a potential adversary's attack or exploitation capabilities against an enterprise's security posture. The Red Team's objective is to improve enterprise cybersecurity by demonstrating the impacts of successful attacks and by demonstrating what works for the defenders (i.e., the Blue Team) in an operational environment. Also known as Cyber Red Team.*[77]

(Continued)

TABLE A.1 (*Continued*)
Glossary of Terms

Term	Definition
Reference Architecture	*A document or set of documents that provides recommended structures and integrations of IT products and services to form a solution. The reference architecture embodies accepted industry best practices, typically suggesting the optimal delivery method for specific technologies. A reference architecture offers IT best practices in an easy-to-understand format that guides the implementation of complex technology solutions.*[78]
Representation Learning	*The success of machine learning algorithms generally depends on data representation, and we hypothesize that this is because different representations can entangle and hide more or less the different explanatory factors of variation behind the data.*[79]
Research	*Research is a systematic and disciplined inquiry that aims to discover, interpret, and expand knowledge in a specific field of study. It is a process of investigation that goes beyond mere observation or gathering of information. Research involves formulating research questions or hypotheses, designing methodologies, collecting and analyzing data, and drawing meaningful conclusions.*[80]
Retrieval Augmented Generation (RAG)	*An architecture for optimizing the performance of an artificial intelligence (AI) model by connecting it with external knowledge bases. RAG helps large language models (LLMs) deliver more relevant responses at a higher quality.*[81]
Sandbox	*A system that allows an untrusted application to run in a highly controlled environment where the application's permissions are restricted to an essential set of computer permissions. In particular, an application in a sandbox is usually restricted from accessing the file system or the network.*[82]
(Scale)-Up	*To increase the size, amount, or importance of something, usually an organization or process.*[83]
Scientific Consensus	*The number of findings that people believe and find completely unremarkable exceeds by orders of magnitude the number in which they end up deeply polarized.*[84] However, this does not mean the majority is correct.
Scientific Management, The Principles	*The central theme of scientific management is the need to substitute industrial harmony and trust for warfare and fear in the workplace. This harmony is accomplished by the following management practices: To gauge market trends and regulate operations in a manner that maximizes capital investment, Sustain the enterprise to assure continuous operation and employment, A continuous balanced operations whereby by personnel as well as the enterprise receive greater economic value, Develop a socially beneficial and healthier conditions of work, Provide conditions for self-expression and self-realization among workers, and Promote a common understanding, tolerance and team spirit.*[85]
Scientific Method	*A systematic process involving steps like defining questions, forming hypotheses, conducting experiments, and analyzing data. It minimizes biases and enables replicable research, leading to groundbreaking discoveries like Einstein's theory of relativity, penicillin, and the structure of DNA. This ongoing approach promotes reason, evidence, and the pursuit of truth in science.*[86]

(*Continued*)

TABLE A.1 (*Continued*)
Glossary of Terms

Term	Definition
Seemingly Conscious AI (SCAI)	*Models that "imitate consciousness in such a convincing way that it would be indistinguishable from a claim that you or I might make to one another about our own consciousness."*[87]
Simultaneous Equations	*A set of two or more equations, each containing two or more variables whose values can simultaneously satisfy both or all the equations in the set, the number of variables being equal to or less than the number of equations in the set.*[88]
Small Language Model	*AI models capable of processing, understanding and generating natural language content. As their name implies, SLMs are smaller in scale and scope than large language models (LLMs).*[89]
Software Singularity	*A bold new vision for the future of enterprise IT—a unified, AI-driven platform that acts as a true co-pilot for every level of an organization. By integrating modular microservices, robust security measures, semantic ontologies, and adaptive cloud-native infrastructure, this system aims to revolutionize how businesses operate.*[90]
Strong Bond Governance	*Governance model with the following attributes.* • *Direct, defined relationships that enables open and valid information between governance members.* • *Led by authorities who are closely connected and strongly bonded.* • *Strong Governance, Risk, and Compliance (GRC) system.* • *Back office and field processes combined into a single information model.* • *Designed for application and use in Mission-Critical Environments.*[91]
Structural Dynamics	*The morphology or patterns of motion towards process equilibrium of interpersonal systems.*[92] Also, see Latent Variables and Torque Clustering.
Structural Equation Model (SEM)	*A statistical model that combines principles of factor analysis and path analysis to represent hypothesized relationships among latent constructs and their observed indicators. It consists of a measurement model, which describes the measurement properties of the indicators, and a structural model, which specifies the causal relationships among the latent variables. SEMs are used to allocate explained and unexplained variance of dependent constructs and estimate the parameters of the model using empirical data.*[93]
Sun Tzu (The Art of War)	*"The art of war is of vital importance to the state. It is a matter of life and death, a road either to safety or to ruin. Hence it is a subject of inquiry which can on no account be neglected." So begins The Art of War, a meditation on the rules of war that was first published in China. Historians don't know the exact date of the book's publication (though they believe it to be in the 4th or 5th century); in fact, they don't even know who wrote it! Scholars have long believed that The Art of War's author was a Chinese military leader named Sun Tzu, or Sunzi. Today, however, many people think that there was no Sun Tzu: Instead, they argue, the book is a compilation of generations of Chinese theories and teachings on military strategy.*[94]

TABLE A.1 (*Continued*)
Glossary of Terms

Term	Definition
Supervised Learning	*A machine learning technique that uses labeled datasets to train artificial intelligence algorithm models to identify the underlying patterns and relationships between input features and outputs. The goal of the learning process is to create a model that can predict correct outputs on new real-world data.*[95]
Synthetic Biology	*A field of science that involves redesigning organisms for useful purposes by engineering them to have new abilities. Synthetic biology researchers and companies around the world are harnessing the power of nature to solve problems in medicine, manufacturing and agriculture.*[96]
Taxonomy	*Practice and science concerned with classification or categorization. Typically, there are two parts to it: the development of an underlying scheme of classes (a taxonomy) and the allocation of things to the classes (classification).*[97]
Temperature	*In the context of AI language models, temperature is a hyperparameter that controls the randomness and creativity of the model's output. A higher temperature results in more diverse and unpredictable responses, while a lower temperature produces more focused and deterministic outputs.*[98]
Tokens and Tokenization	*A token is a collection of characters that has semantic meaning for a model. Tokenization is the process of converting the words in your prompt into tokens.*[99]
Torque Clustering	*Designed to efficiently analyze large datasets across various fields, including biology, chemistry, astronomy, psychology, finance, and medicine. By uncovering hidden patterns, it can provide valuable insights, such as detecting disease trends, identifying fraudulent activities, and understanding human behavior.*[100]
Transformer Neural Network	*The transformer is a component used in many neural network designs for processing sequential data, such as natural language text, genome sequences, sound signals or time series data. Most applications of transformer neural networks are in the area of natural language processing. A transformer neural network can take an input sentence in the form of a sequence of vectors, and converts it into a vector called an encoding, and then decodes it back into another sequence. An important part of the transformer is the attention mechanism. The attention mechanism represents how important other tokens in an input are for the encoding of a given token.*[101]
Vibe Computing	*A way of building software where you tell an AI what you want in plain language, and the AI writes the code for you. Instead of sitting down and typing every line by hand, you describe the functionality or problem you're trying to solve, and AI handles the technical side. Vibe coding feels more like giving directions.*[102]
Zero-Shot Learning	*A machine learning scenario in which an AI model is trained to recognize and categorize objects or concepts without having seen any examples of those categories or concepts beforehand.*[103]

The National Institute of Standards and Technology (NIST) has two 'unofficial' glossaries of possible interest. Their author indicates that these glossaries are subject to change:

- *An aggregation of terms and definitions specified in NIST's cybersecurity and privacy standards, guidelines, and other technical publications, and in CNSSI 4009.*[104]
- The Language of Trustworthy AI: An In-Depth Glossary of Terms.[105]

Other AI Glossaries are emerging, an example of which follows.[106]

- **Automatic Prompt Generation**: A system-level process using AI to discover and refine effective prompts.
- **Calibrated Confidence Prompting (CCP)**: Requiring a model to state its confidence level for factual claims.
- **Chiaroscuro**: The use of strong contrasts between light and dark in art.
- **Clarification Prompting**: Asking a model to define its terms or resolve ambiguities.
- **Cognitive Survival**: The ability to think clearly and make effective decisions amidst overwhelming complexity.
- **Complexity Compression**: Transforming an overwhelming problem into something testable or explorable.
- **Context-Aware Decomposition (CAD)**: Breaking a complex problem into components, solving each, and synthesizing the solutions.
- **Context Window**: The short-term memory of a language model for a specific interaction.
- **Controlled Hallucination for Ideation (CHI)**: Intentionally using a model's ability to generate non-existent ideas for creative brainstorming.
- **Counterfactual or Hypothetical Prompting**: Asking a model to explore a 'what if' scenario.
- **Desocialization**: The loss of social and communication skills due to algorithm-driven digital environments.
- **Error-Guided Prompting**: Providing direct feedback on an error and asking the model to correct it.
- **Few-Shot Prompting**: Providing a model with multiple examples to demonstrate a pattern for the desired output.
- **Generative AI**: A class of AI systems that can create new content like text, images, or code.
- **Hallucinated Information**: False information generated by a model when it lacks necessary data and predicts text that seems plausible.
- **High-Agency Questions**: Questions that compress complexity, generate useful outputs, and are reusable.
- **Large Language Model (LLM)**: A type of neural network trained on vast amounts of text to predict subsequent words in a sequence.
- **Latent Personality**: The underlying behavior, tone, and ethical guardrails of an AI model.

- **Many-Shot Learning**: Using a large number of examples in a prompt to achieve high performance on complex tasks.
- **Mechanized Patterns of Inquiry**: Questions that feel robotic and lack genuine human curiosity.
- **Meta-Prompting**: Asking an AI to act as an expert to help create a better prompt for a specific goal.
- **Mnemonic**: A memory aid, such as a pattern or phrase.
- **Model-Agnostic**: A method or principle that is effective across different types of models.
- **Multi-Perspective Simulation (MPS)**: Simulating a dialogue between different expert personas to analyze an issue.
- **One-Shot Prompting**: Providing a single example to guide a model's format or tone.
- **Output Generation**: The ability of a question directly leads to decisions, actions, or frameworks.
- **Progressive Prompting**: Guiding a model through a topic by gradually increasing complexity.
- **Prompt Chaining**: Breaking a complex task into a series of interconnected prompts.
- **Prompt Injection**: Malicious inputs designed to trick an AI into bypassing its safety protocols.
- **Prompt or Perish**: A guiding motto that stresses the increasing importance of prompt literacy.
- **Psychological friction**: Mental resistance or discomfort caused by poorly framed questions.
- **Recursive Self-Improvement Prompting (RSIP)**: A process where a model generates content, critiques it, and refines it iteratively.
- **Reflection Prompting**: Asking a model to evaluate its own output to identify flaws.
- **Retrieval-Augmented Generation (RAG)**: A technique enabling a model to retrieve information from an external source before responding.
- **Reusability Factor**: A quality of a question that allows it to be used in different situations to produce clarity.
- **Role Prompting**: Instructing an AI to adopt a specific persona.
- **RTF (Role-Task-Format) Framework**: A three-component structure for prompts that sets the AI's persona, its specific task, and the desired output format.
- **Self-Consistency Decoding**: Generating multiple reasoning paths and selecting the most common answer.
- **Semi Skill**: An ability requiring both the nuance of soft skills and the precision of hard skills.
- **Signal to noise ratio**: A concept describing how much useful information a question produces versus distracting information.
- **Step-Back Prompting**: Guiding an AI to think about abstract concepts before tackling a specific question.

- **Structured Inputs**: Carefully designed prompts that provide a framework for useful outputs.
- **System Prompt**: A set of foundational instructions given to an LLM that defines its identity, rules, and capabilities.
- **The Fog (of Ambiguity)**: A metaphor for mental confusion or cognitive overwhelm that impairs decision-making.
- **Tokenization**: The process where a model breaks down instructions into smaller units (tokens).
- **Tracy Tone**: A framework used to control the voice, behavior, and style of an AI.
- **Tree-of-Thought Prompting**: An advanced technique where a model explores multiple parallel lines of reasoning.
- **Zero-Shot Prompting**: Giving a model a direct instruction without any prior examples.

Executives should be cautious when using materials from third parties not affiliated with Bonafide research institutions and/or universities, and commercial firms. Not all providers subject their documentation to vigorous academic style rigor, and the possibility of confusion may exist. A common vocabulary is a critical component of Operational Excellence.

NOTES

1. "Adoption Readiness Levels (ARL) Framework." U.S. Department of Energy. https://www.energy.gov/technologytransitions/adoption-readiness-levels-arl-framework?nrg_redirect=466798. Accessed April 6, 2025.
2. "Advanced Persistent Threats (APT) Explained." CrowdStrike. https://www.crowdstrike.com/en-us/cybersecurity-101/threat-intelligence/advanced-persistent-threat-apt/. Accessed August 21, 2025.
3. "What Is Agentic AI, and How Will It Change Work?" Harvard Business Review. https://hbr.org/2024/12/what-is-agentic-ai-and-how-will-it-change-work. Accessed March 31, 2025.
4. "AI Detection." AI-Tool.ai. https://ai-tool.ai/ai-glossary/fundamentals/ai-detection. Accessed May 28, 2025.
5. "What Is AI Ethics? IBM. https://www.ibm.com/think/topics/ai-ethics. Accessed May 28, 2025.
6. "What Are AI Hallucinations?" IBM. https://www.ibm.com/think/topics/ai-hallucinations. Accessed February 23, 2025.
7. "What Is an Algorithm? | Definition & Examples." Scribbr. https://www.scribbr.com/ai-tools/what-is-an-algorithm/. Accessed February 23, 2025.
8. "API." TechTerms.com. https://techterms.com/definition/api#google_vignette. Accessed April 30, 2025.
9. "Bayesian Optimization – What Is It? How to Use It Best?" Inside Machine Learning. Accessed September 4, 2025.
10. "Bias." Psychology Today. https://www.psychologytoday.com/us/basics/bias?msockid=39454459d22269f52e3f51a3d36b68ad. Accessed May 27, 2025.
11. "bionics." Britannica. https://www.britannica.com/technology/bionics. Accessed April 13, 2025.
12. "What Is a Black Swan Event?" CFI. https://corporatefinanceinstitute.com/resources/economics/black-swan-event/. Accessed August 7, 2025.

13. "Emerging Technologies for Inclusive Learning." Learning Guild. https://www. learningguild.com/insights/315/emerging-technologies-for-inclusive-learning/?utm_ campaign=32565442-Learning%20%7C%20US%20%7C%20Learning%20Guild%20 2025&utm_source=email&utm_medium=email&utm_term=Research-ET25&utm_ content=Research-ET25_EmailPromo_250227. Accessed February 28, 2025.
14. "Bloom's Taxonomy of Learning." SimplyPsychology. https://www.simplypsychology. org/blooms-taxonomy.html. Accessed March 8, 2025.
15. "Brain Machine Interface." Stanford University. https://colemanlab.stanford.edu/research/ brain-machine-interface. Accessed September 7, 2025.
16. "What is Chain of Thoughts (CoT)?" IBM. https://www.ibm.com/think/topics/chain-of-thoughts. Accessed February 26, 2025.
17. "Checksum." TechTerms.com. https://techterms.com/definition/checksum. Accessed August 6, 2025.
18. "What is a context window?" IBM. https://www.ibm.com/think/topics/context-window. Accessed March 8, 2025.
19. "What Is Contrastive Learning?" CLRN. https://www.clrn.org/what-is-contrastive-learning/. Accessed April 23, 2025.
20. "Critical Thinking." Merriam-Webster. https://www.merriam-webster.com/dictionary/ critical%20thinking. Accessed March 8, 2025.
21. "What Is Cross-Selling? A Complete Guide." Salesforce. https://www.salesforce.com/ sales/cross-selling/. Accessed April 12, 2025.
22. "Cultural Cognition and Scientific Consensus." Yale Scientific. https://www. yalescientific.org/2011/05/cultural-cognition-and-scientific-consensus/. Accessed April 10, 2025.
23. "Culture." Merriam-Webster.com Dictionary, Merriam-Webster, https://www.merriam-webster.com/dictionary/culture. Accessed February 23, 2025.
24. "What Is the Curse of Dimensionality?" builtin. https://builtin.com/data-science/curse-dimensionality. Accessed October 4, 2025.
25. "Dark Energy." Britannica. https://www.britannica.com/science/dark-energy. Accessed March 7, 2025.
26. "What Is Dark Matter?" SPACE.com. https://www.space.com/20930-dark-matter.html. Accessed March 7, 2025.
27. "What Is Multi Modal Data?" CLRN. https://www.clrn.org/what-is-multi-modal-data/ #google_vignette. Accessed May 2, 2025.
28. "What Is Data Validation?" TechTarget. https://www.techtarget.com/searchdata management/definition/data-validation. Accessed March 8, 2025.
29. "Deep Learning Architectures." IBM. https://developer.ibm.com/articles/cc-machine-learning-deep-learning-architectures/. Accessed February 23, 2025.
30. "Deep Neural Network: What Is It and How Is It Working?" DataScientest. https:// datascientest.com/en/deep-neural-network-what-is-it-and-how-is-it-working. Accessed February 23, 2025.
31. "AI 'Urania' Transforms Gravitational Wave Detection with Revolutionary Designs." OPENTOOLS. https://opentools.ai/news/ai-urania-transforms-gravitational-wave-detection-with-revolutionary-designs. Accessed April 20, 2025.
32. "Difference Equations." LibreTexts: Mathematics. https://math.libretexts.org/ Bookshelves/Analysis/Supplemental_Modules_(Analysis)/Ordinary_Differential_ Equations/2%3A_First_Order_Differential_Equations/2.1%3A_Difference_ Equations. Accessed March 10, 2025.
33. "What Are Diffusion Models." IBM. https://www.ibm.com/think/topics/diffusion-models. Accessed March 8, 2025.
34. "Dormitive Principle." WordSense. https://www.wordsense.eu/dormitive_principle/. Accessed August 8, 2025.
35. "Dunning–Kruger Effect." Wikipedia. https://en.wikipedia.org/wiki/Dunning%E2% 80%93Kruger_effect. April 10, 2025.

36. "How NASA Is Testing AI to Make Earth-Observing Satellites Smarter." NASA. https://www.nasa.gov/science-research/earth-science/how-nasa-is-testing-ai-to-make-earth-observing-satellites-smarter/?utm_source=Generative_AI&utm_medium=Newsletter&utm_campaign=meta-names-ex-openai-scientist-to-head-superintelligence-labs&_bhlid=4173a4c020e532434df54d1d9ae3d301fe8888ff. Accessed July 28, 2025.

37. "What Is the Eisenhower Matrix? Definition, Origin, Method, Case Use, Examples and More." IDEASCALE. https://ideascale.com/blog/eisenhower-matrix-a-guide/. Accessed April 16, 2025.

38. "What Are Embeddings in Machine Learning?" geeksforgeeks. https://www.geeksforgeeks.org/what-are-embeddings-in-machine-learning-2/. Accessed February 23, 2025.

39. "Expert System." TechTarget. https://www.techtarget.com/searchenterpriseai/definition/expert-system. Accessed April 11, 2025.

40. "What Is Fine-Tuning?" IBM. https://www.ibm.com/think/topics/fine-tuning. Accessed March 8, 2025.

41. "What Is FOBO? How to Stay Relevant and Not Become Obsolete at Work." Forbes. https://www.forbes.com/sites/dianehamilton/2025/03/03/what-is-fobo-how-to-stay-relevant-and-not-become-obsolete-at-work/. Accessed April 10, 2025.

42. "Future-proof." Collinsdictionary.com. Collins. Accessed February 23, 2025. https://www.collinsdictionary.com/us/dictionary/english/future-proof.

43. "What's Generative Engine Optimization (GEO) & How to Do It." Foundation. https://foundationinc.co/. Accessed May 5, 2025.

44. "What Is a GPU and Its Role in AI?" Google Cloud. https://cloud.google.com/discover/gpu-for-ai. Accessed June 3, 2025.

45. "Heuristic." Merriam-Webster. https://www.merriam-webster.com/dictionary/heuristic. Accessed October 8, 2025.

46. "Independent Verification & Validation (IV&V)." NIST. https://csrc.nist.gov/glossary/term/independent_verification_and_validation. Accessed June 5, 2025.

47. "Unlock Inference-as-a-Service with Cloud Run and Vertex AI." Google Cloud. https://cloud.google.com/blog/products/ai-machine-learning/improve-your-gen-ai-app-velocity-with-inference-as-a-service. Accessed March 17, 2025.

48. "Inference Engine." ScienceDirect. https://www.sciencedirect.com/topics/computer-science/inference-engines. Accessed April 11, 2025.

49. "Understanding Inference-Time Scaling." Medium. https://medium.com/predict/understanding-inference-time-scaling-14a2501a1265. Access July 3, 2025.

50. "What Is Instant Gratification? (Definition & Examples)." PositivePsychology.com https://positivepsychology.com/instant-gratification/. Accessed April 30, 2025.

51. "JEPA." geeksforgeeks. https://www.geeksforgeeks.org/artificial-intelligence/jepa/. Accessed August 6, 2025.

52. "Key Performance Indicators (KPIs)." Investopedia. https://www.investopedia.com/terms/k/kpi.asp. Accessed February 22, 2025.

53. "What Is a Knowledge Graph?" geeksforgeeks. https://www.geeksforgeeks.org/data-analysis/what-is-a-knowledge-graph/. Accessed August 6, 2025.

54. "Defining 'Knowledge' in a Purchase Agreement." Winston & Strawn. https://www.winston.com/en/insights-news/defining-knowledge-in-a-purchase-agreement. Accessed February 23, 2025.

55. "What Are large language models (LLMs)?" IBM. https://www.ibm.com/think/topics/large-language-models. Accessed February 26, 2025.

56. "How Reasoning Models Are Transforming Logical AI Thinking." Microsoft Developer Community Blog. https://techcommunity.microsoft.com/blog/azuredevcommunityblog/how-reasoning-models-are-transforming-logical-ai-thinking/4373194. Accessed April 2, 2025.

57. "What Is Latent Variable?" geeksforgeeks. https://www.geeksforgeeks.org/what-is-latent-variable/. Accessed February 26, 2025.

58. "Unintended Consequences." Econlib. https://www.econlib.org/library/Enc/Unintended Consequences.html. Accessed March 7, 2025.

59. "Methods & Designs: A guide to LISREL-Type Structural Equation Modeling." Behavior Research Methods & Instrumentation. https://link.springer.com/content/pdf/10.3758/BF03202105.pdf. Accessed March 8, 2025.
60. "Marginal Cost." Economics Online. https://www.economicsonline.co.uk/definitions/marginal_cost.html/. Accessed April 5, 2025.
61. "Meta Prompting." geeksforgeeks. https://www.geeksforgeeks.org/artificial-intelligence/meta-prompting/. Accessed August 5, 2025.
62. "What Is Model Collapse?" IBM. https://www.ibm.com/think/topics/model-collapse. Accessed June 6, 2025.
63. "What Is Model Context Protocol (MCP)?" IBM. https://www.ibm.com/think/topics/model-context-protocol. Accessed July 20, 2025.
64. "Moral Foundations Theory." MoralFoundations.org. https://moralfoundations.org/. Accessed June 7, 2025.
65. "Morphogenesis." Britannica. https://www.britannica.com/science/morphogenesis. Accessed April 21, 2025.
66. "What Is NLP?" IBM. https://www.ibm.com/think/topics/natural-language-processing. Accessed June 17, 2025.
67. "5 Steps to Define and Achieve Your North Star Goal." Statsig. https://www.statsig.com/perspectives/steps-to-define-and-achieve-your-north-star-goal. Accessed July 25, 2025.
68. "Introduction to Numerical Relativity." Cornell University. https://arxiv.org/abs/2008.12931. Accessed April 20, 2025.
69. "Open Source." Wikipedia. https://en.wikipedia.org/wiki/Open_source. Accessed February 23, 2025.
70. "What Is Pareto Efficiency?" Investopedia. https://www.investopedia.com/terms/p/pareto-efficiency.asp. Accessed March 31, 2025.
71. "What Is a Physical Symbol System." Umb.edu. https://faculty.umb.edu/gary_zabel/Courses/Bodies,%20Souls,%20and%20Robots/Texts/What%20is%20a%20physical%20symbol%20system.htm. Accessed April 8, 2025.
72. "What Is Predictive AI?" IBM. https://www.ibm.com/think/topics/predictive-ai. Accessed May 5, 2025.
73. "What Is Prompt Chaining?" IBM. https://www.ibm.com/think/topics/prompt-chaining. Accessed February 26, 2025.
74. "What Is Prompt Engineering?" McKinsey & Company. https://www.mckinsey.com/featured-insights/mckinsey-explainers/what-is-prompt-engineering. Accessed February 23, 2025.
75. "Get Started with Prompt Library." Microsoft. https://learn.microsoft.com/en-us/ai-builder/prompt-library. Accessed August 5, 2025.
76. "What Is Quantum Computing?" IBM. https://www.ibm.com/think/topics/quantum-computing. Accessed March 1, 2025.
77. "Red Team." NIST. https://csrc.nist.gov/glossary/term/Red_Team. Accessed May 1, 2025.
78. "What Is a Reference Architecture?" Hewlett Packard Enterprise. https://www.hpe.com/us/en/what-is/reference-architecture.html. Accessed April 30, 2025.
79. "Representation Learning: A Review and New Perspectives." Princeton. https://www.cs.princeton.edu/courses/archive/spring13/cos598C/Representation%20Learning%20-%20A%20Review%20and%20New%20Perspectives.pdf. Accessed February 23, 2025.
80. "What Is Research? – Definition, Objectives & Types of Research." LIS Education Network. https://www.lisedunetwork.com/what-is-research/#google_vignette. Accessed April 10, 2025.
81. "What Is RAG (Retrieval Augmented Generation)?" IBM. https://www.ibm.com/think/topics/retrieval-augmented-generation. Accessed February 26, 2025.
82. "Sandbox." NIST. https://csrc.nist.gov/glossary/term/Sandbox. Accessed April 9, 2025.
83. "Scale Something Up." Cambridge Dictionary. https://dictionary.cambridge.org/dictionary/english/scale-up. Accessed April 10, 2025.
84. "Cultural Cognition and Scientific Consensus." Yale Scientific. https://www.yalescientific.org/2011/05/cultural-cognition-and-scientific-consensus/. Accessed April 10, 2025.

85. "Scientific Management and the Knowledge Worker of the 1990's." Proceedings of the 11th Annual Association of Management Conference. https://therrinstitute.com/wp-content/uploads/2019/10/1993-Scientific-Management-and-the-Knowledge-Worker-of-the-1990s.pdf. Accessed April 7, 2025.

86. "What Is the Scientific Method: How Does It Work and Why Is It Important?" AJE. https://www.aje.com/arc/what-is-the-scientific-method/. Accessed February 23, 2025.

87. "Microsoft AI CEO Suleyman is worried about 'AI psychosis' and AI that seems 'conscious'." Fortune. https://fortune.com/2025/08/22/microsoft-ai-ceo-suleyman-is-worried-about-ai-psychosis-and-seemingly-conscious-ai/. Accessed August 30, 2025.

88. "Simultaneous Equations." Dictionary.com. https://www.dictionary.com/browse/simultaneous-equations. Accessed March 11, 2025.

89. "What Are Small Language Models?" IBM. https://www.ibm.com/think/topics/small-language-models. Accessed July 18, 2025.

90. "Software Singularity: Toward a Unified AI-Driven Enterprise Co-Pilot." VaidhyaMegha. https://www.linkedin.com/pulse/enterprise-software-singularity-vaidhyamegha-cmumc/. Accessed April 22, 2025.

91. Holland, Winford "Dutch" E., and Shemwell, Scott M. (2014). *Implementing a Culture of Safety: A Roadmap to Performance-Based Compliance.* New York: Xlibris. https://www.xlibris.com/Bookstore/BookDetail.aspx?BookId=SKU-0143303003

92. Shemwell, Scott M. (2015). *Structural Dynamics: Foundation of Next Generation Management Science.* Houston: RRI Publications. http://www.amazon.com/Structural-Dynamics-Foundation-Generation-Management-ebook/dp/B00U0JKMT0

93. "Structural Equation Model." ScienceDirect. https://www.sciencedirect.com/topics/computer-science/structural-equation-model. Accessed February 27, 2025.

94. "The Art of War." History. https://www.history.com/articles/the-art-of-war. Access April 5, 2025.

95. "What Is Supervised Learning?" IBM. https://www.ibm.com/think/topics/supervised-learning. Accessed February 27, 2025.

96. "Synthetic Biology." National Human Genome Research Institute. https://www.genome.gov/about-genomics/policy-issues/Synthetic-Biology. Accessed April 21, 2025.

97. "Taxonomy." Wikipedia. https://en.wikipedia.org/wiki/Taxonomy. Accessed March 19, 2025.

98. "Temperature." Full Stack AI. https://fullstackai.co/ai-glossary/temperature/. Accessed March 8, 2025.

99. "Tokens and Tokenization." IBM. https://www.ibm.com/docs/en/watsonx/saas?topic=solutions-tokens. Accessed March 8, 2025.

100. "Scientists Unveil AI That Learns without Human Labels – A Major Leap toward True Intelligence!" SciTechDaily. https://scitechdaily.com/scientists-unveil-ai-that-learns-without-human-labels-a-major-leap-toward-true-intelligence/. Accessed February 23, 2025.

101. "Transformer Neural Network." DeepAI. https://deepai.org/machine-learning-glossary-and-terms/transformer-neural-network. Accessed March 8, 2025.

102. "What Is Vibe Coding? Definition, Tools, Pros, and Cons." Datacamp. https://www.datacamp.com/blog/vibe-coding. Accessed March 8, 2025.

103. "What Is Zero-Shot Learning?" IBM. https://www.ibm.com/think/topics/zero-shot-learning. Accessed March 8, 2025.

104. "Glossary." NIST. https://csrc.nist.gov/glossary. Accessed April 9, 2025.

105. "The Language of Trustworthy AI: An In-Depth Glossary of Terms." NIST. https://airc.nist.gov/glossary/. Accessed April 9, 2025.

106. "Learn to Prompt (2025 Guide)." Max Berry. https://www.maxberry.ca/p/learn-to-prompt-2025-guide?img=https%3A%2F%2Fsubstack-post-media.s3.amazonaws.com%2Fpublic%2Fimages%2F942dceae-57fa-4f8c-be69-875fe76745a6_1024x1024.png&open=false. Accessed July 15, 2025.

Appendix II
Variable Types

As with the Appendix I: Glossary of Terms, the following terms are used in this report and by practitioners. These definitions are provided for the reader's convenience. They can also be used in Key Word searches. As used throughout this book, direct quoted text is in *italics*, per editorial convention.

The following list of 27 different types of variables used in research and statistics are of particular interest to those developing, interpreting results, and generally using the output from AI solutions.[1]

Note: Some variables are individually cited and while each citation is not included for space reasons, the overall reference documents each item. This is consistent with best citation practices.

TABLE A.2
Types of Variables

Variable	Definition
Quantitative (Numerical)	*Quantifiable in nature and represented in numbers, allowing the data collected to be measured on a scale or range. These variables generally yield data that can be organized, ranked, measured, and subjected to mathematical operations.*
Continuous	*Subtype of quantitative variables that can have an infinite number of measurements within a specified range. They provide detailed insights based on precise measurements and are often representative on a continuous scale.*
Discrete	*A form of quantitative variable that can only assume a finite number of values. They are typically count-based.*
Qualitative (Categorical)	*Non-numerical data points that categorize or group data entities based on shared features or qualities.*
Nominal	*Subtype of qualitative variables represent categories without any inherent order or ranking.*
Ordinal	*Subtype of categorical (qualitative) variables with a key feature of having a clear, distinct, and meaningful order or ranking to the categories.*
Dichotomous (Binary)	*Type of categorical variable that consist of only two opposing categories like true/false, yes/no, success/failure, and so on.*
Ratio	*The highest level of quantitative variables that contain a zero point or absolute zero, represents a complete absence of the quantity.*
Interval	*Quantitative variables that have equal, predictable differences between values, but they do not have a true zero point.*
Dependent	*The outcome or effect that the researcher wants to study. Its value depends on or is influenced by one or more other variables known as independent variables.*

(Continued)

TABLE A.2 (*Continued*)
Types of Variables

Variable	Definition
Independent	*Presumed to have some effect on the dependent variable in a study. It can often be thought of as the cause in a cause-and-effect relationship.*
Confounding (Confounders)	*Variables that might distort, confuse or interfere with the relationship between an independent variable and a dependent variable, leading to a false correlation.*
Control	*Variables in a research study that the researcher keeps constant to prevent them from interfering with the relationship between independent and dependent variables.*
Latent	*Also referred to as hidden or unobserved variables—are variables that are not directly observed or measured but are inferred from other variables that are observed (measured directly).*
Derived	*Variables that are created or developed based on existing variables in a dataset. They involve applying certain calculations or manipulations to one or more variables to create a new one.*
Time-Series	*Variables are a set of data points ordered or indexed in time order. They provide a sequence of data points, each associated with a specific instance in time.*
Cross-Sectional	*Data collected from many subjects at the same point in time or without regard to differences in time.*
Predictor	*Also known as independent or explanatory variable—is a variable that is being manipulated in an experiment or study to see how it influences the dependent or response variable.*
Response	*Also known as the dependent or outcome variable—is what the researcher observes for any changes in an experiment or study. Its value depends on the predictor or independent variable.*
Exogenous	*Variables that are not affected by other variables in the system but can affect other variables within the same system.*
Endogenous	*Variables whose value is determined by the functional relationships within the system in an economic or statistical model. They depend on the values of other variables in the model.*
Causal	*Variables which can directly cause an effect on the outcome or dependent variable. Their value or level determines the value or level of other variables.*
Moderator	*Variables that can affect the strength or direction of the association between the predictor (independent) and response (dependent) variable. They specify when or under what conditions a relationship holds.*
Mediator	*Variables that account for, or explain, the relationship between an independent variable and a dependent variable, providing an understanding of 'why' or 'how' an effect occurs.*
Extraneous	*Variables that are not of primary interest to a researcher but might influence the outcome of a study. They can add 'noise' to the research data if not controlled.*
Dummy	*Often used in regression analysis, are artificial variables created to represent an attribute with two or more distinct categories or levels.*
Composite	*New variables created by combining or grouping two or more variables.*

Additional variable types (including new ones) may be used in AI solutions. However, this list captures those main and frequently used expressions.

NOTE

1. "27 Types of Variables in Research and Statistics." HelpfulProfessor.com. https://helpfulprofessor.com/types-of-variables/. Accessed March 18, 2025.

Appendix III
Major AI Resources

The following is a partial list of available AI resources, both public and private. Its purpose is to provide readers with an initial starting point to find AI materials that may be useful to their organization. This field is changing rapidly, and readers may find more appropriate (for fee/free) solutions from these and other providers.

This information is provided for discussion and educational purposes and neither the author nor the publisher takes any responsibility for its accuracy or appropriateness.

As used throughout this book, direct quoted text is in *italics*, per editorial convention.

TABLE A.3
Major AI Resources

Resource	Overview
AI.gov	*The United States is in a race to achieve global dominance in artificial intelligence. Whoever has the largest AI ecosystem will set the global standards and reap broad economic and security benefits.*[1]
AI-Tool.ai	*AI tools directory. AI tools list & GPTs store are updated daily by ChatGPT.* Very comprehensive and information as well as update constantly. An excellent reference.[2]
America's AI Action Plan	*America must have the most powerful AI systems in the world, but we must also lead the world in creative and transformative application of these systems. Achieving these goals requires the Federal government to create the conditions where private-sector-led innovation can flourish.*[3]
Anthropic	*An AI safety and research company. We build reliable, interpretable, and steerable AI systems.*[4]
Center for AI Standards and Innovation (CAISI)	*CAISI will serve as industry's primary point of contact within the U.S. Government to facilitate testing and collaborative research related to harnessing and securing the potential of commercial AI systems.*[5]
Cybersecurity Maturity Model Certification 2.0 Program	*The next iteration of the CMMC cybersecurity model. It streamlines requirements to three levels of cybersecurity and aligns the requirements at each level with well-known and widely accepted NIST cybersecurity standards.*[6]
Developer Roadmaps	*A community effort to create roadmaps, guides and other educational content to help guide developers in picking up a path and guide their learnings.*[7]

(Continued)

TABLE A.3 (*Continued*)
Major AI Resources

Resource	Overview
FinOps Foundation	*Project of The Linux Foundation (alongside organizations like Cloud Native Computing Foundation) dedicated to advancing people who practice the discipline of cloud financial management through best practices, education, and standards. The FinOps Foundation is a 60,000+ person-strong community, representing more than 15,000+ companies.*[8]
Fintech Open Source Foundation	*An umbrella organization under the Linux Foundation, whose purpose is to accelerate collaboration and innovation in financial services through the adoption of open source software, standards and best practices.*[9]
IEEE	*The world's largest technical professional organization dedicated to advancing technology for the benefit of humanity. IEEE and its members inspire a global community through its highly cited publications, conferences, technology standards, and professional and educational activities.*[10]
ISO/IEC 42001:2023	*An international standard that specifies requirements for establishing, implementing, maintaining, and continually improving an Artificial Intelligence Management System (AIMS) within organizations. It is designed for entities providing or utilizing AI-based products or services, ensuring responsible development and use of AI systems.*[11]
Project Management Institute (PMI)	*The leading authority in project management, dedicated to guiding the way to project success. Since 1969, PMI has shone a light on the power of project management and the people behind the projects.*[12]
Society for Human Resource Management (SHRM)	This is a well-known industry group, but it also has a significant Body of Knowledge about AI in the Workplace available to non-members.[13]
Taxonomy and Definitions for Terms Related to Driving Automation Systems for On-Road Motor Vehicles	*A taxonomy with detailed definitions for six levels of driving automation.*[14]
The Committee of Sponsoring Organizations of the Treadway Commission (COSO)	*The mission is to help organizations improve performance by developing thought leadership that enhances internal control, risk management, governance and fraud deterrence.*[15]

NOTES

1. "AI GOV." The White House. https://www.ai.gov/. Accessed July 24, 2025.
2. "Discover the Best AI Websites & Tools." AI-Tool.ai. https://ai-tool.ai/. Accessed May 28, 2025.
3. "*Winning the Race:* America's AI Action Plan." The White House. https://www.whitehouse.gov/wp-content/uploads/2025/07/Americas-AI-Action-Plan.pdf?utm_source=Generative_AI&utm_medium=Newsletter&utm_campaign=google-deepmind-decodes-the-past-by-training-on-ancient-texts&_bhlid=3f842c21ab35b1342f91d98e5f3d94e2bc7485ff. Accessed July 24, 2025.

4. "Anthropic." https://www.anthropic.com/company. Accessed June 9, 2025.
5. "Statement from U.S. Secretary of Commerce Howard Lutnick on Transforming the U.S. AI Safety Institute into the Pro-Innovation, Pro-Science U.S. Center for AI Standards and Innovation." U.S. Department of Commerce. https://www.commerce.gov/news/press-releases/2025/06/statement-us-secretary-commerce-howard-lutnick-transforming-us-ai. Accessed June 6, 2025.
6. "Cybersecurity Maturity Model Certification 2.0 Program." CISA. https://www.cisa.gov/resources-tools/resources/cybersecurity-maturity-model-certification-20-program. Accessed September 10, 2025.
7. "Developer Roadmaps." Insight Media Group. https://roadmap.sh/. Accessed August 5, 2025.
8. "FinOps Foundation." https://www.finops.org/. Accessed June 6, 2025.
9. "Fintech Open Source Foundation." FINOS. https://www.finos.org/?_gl=1*1ysaf9i*_ga*MTM4MDE5NjQ0Ny4xNzUwODU2MDEz*_ga_96R2EHV3BL*czE3NTA4NTYwMTIkbzEkZzEkdDE3NTA4NTYzMDQkajQ5JGwwJGgw. Accessed June 25, 2025.
10. "IEEE." https://www.ieee.org/about/at-a-glance. Accessed July 7, 2025.
11. "ISO/IEC 42001:2023." ISO. https://www.iso.org/standard/42001. Accessed May 12, 2025.
12. "Project Management Institute." https://www.pmi.org/about/our-legacy. Accessed July 7, 2025.
13. "Artificial Intelligence in the Workplace." SHRM. https://www.shrm.org/topics-tools/topics/artificial-intelligence-in-the-workplace. Accessed October 8, 2025.
14. "3016_202104 – Taxonomy and Definitions for Terms Related to Driving Automation Systems for On-Road Motor Vehicles." SAE International. https://www.sae.org/standards/j3016_202104-taxonomy-definitions-terms-related-driving-automation-systems-road-motor-vehicles. Accessed October 4, 2025.
15. "COSO." https://www.coso.org/. Accessed October 10, 2010.

Index

For Product Safety Concerns and Information please contact our EU
representative GPSR@taylorandfrancis.com
Taylor & Francis Verlag GmbH, Kaufingerstraße 24, 80331 München, Germany

www.ingramcontent.com/pod-product-compliance
Lightning Source LLC
Chambersburg PA
CBHW070950200526
45161CB00001BA/63